Representations of $SL_2(\mathbb{F}_q)$

Algebra and Applications

Volume 13

Algebra and Applications aims to publish well written and carefully refereed monographs with up-to-date information about progress in all fields of algebra, its classical impact on commutative and noncommutative algebraic and differential geometry, K-theory and algebraic topology, as well as applications in related domains, such as number theory, homotopy and (co)homology theory, physics and discrete mathematics.

Particular emphasis will be put on state-of-the-art topics such as rings of differential operators, Lie algebras and super-algebras, group rings and algebras, C^*-algebras, Kac-Moody theory, arithmetic algebraic geometry, Hopf algebras and quantum groups, as well as their applications. In addition, Algebra and Applications will also publish monographs dedicated to computational aspects of these topics as well as algebraic and geometric methods in computer science.

Cédric Bonnafé

Representations of $\mathsf{SL}_2(\mathbb{F}_q)$

Springer

Cédric Bonnafé
CNRS (UMR 5149)
Université Montpellier 2
Institut de Mathématiques et de Modélisation de Montpellier
Place Eugène Bataillon
34095 Montpellier Cedex
France
cedric.bonnafe@math.univ-montp2.fr

ISBN 978-1-4471-2599-0 ISBN 978-0-85729-157-8 (eBook)
DOI 10.1007/978-0-85729-157-8
Springer London Dordrecht Heidelberg New York

British Library Cataloguing in Publication Data
A catalogue record for this book is available from the British Library

Mathematics Subject Classification (2010): 20, 20-02

Cover design: VTEX, Vilnius

Printed on acid-free paper

Springer is part of Springer Science+Business Media (www.springer.com)

To my parents.

Preface

If our sole purpose were to calculate the character table of the finite group $G = \mathrm{SL}_2(\mathbb{F}_q)$ (here, q is a power of a prime number p) by ad hoc methods, this book would only amount to a few pages. Indeed, this problem was solved independently by Jordan [Jor] and Schur [Sch] in 1907. The goal of this book is rather to use the group G to give an introduction to the ordinary and modular representation theory of finite reductive groups, and in particular to Harish-Chandra and Deligne-Lusztig theories. It is addressed in particular to students who would like to delve into Deligne-Lusztig theory with a concrete example at hand. The example of $G = \mathrm{SL}_2(\mathbb{F}_q)$ is sufficiently simple to allow a complete description, and yet sufficiently rich to illustrate some of the most delicate aspects of the theory.

There are a number of excellent texts on Deligne-Lusztig theory (see for example Lusztig [Lu1], Carter [Carter], Digne-Michel [DiMi] for the theory of ordinary characters and Cabanes-Enguehard [CaEn] for modular representations). This book does not aim to offer a better approach, but rather to complement the general theory with an illustrated example. We have tried not to rely upon the above books and give full proofs, in the example of the group G, of certain general theorems of Deligne-Lusztig theory (for example the Mackay formula, character formulas, questions of cuspidality etc.). Although it is not always straightforward, we have tried to give proofs which reflect the spirit of the general theory, rather than giving ad hoc arguments. We hope that this shows how general arguments of Deligne-Lusztig theory may be made concrete in a particular case. At the end of the book we have included a chapter offering a very succinct overview (without proof) of Deligne-Lusztig theory in general, as well as making links to what has already been seen (see Chapter 12).

Historically, the example of $\mathrm{SL}_2(\mathbb{F}_q)$ played a seminal role. In 1974, Drinfeld (at age nineteen!) constructed a Langlands correspondence for $\mathrm{GL}_2(\mathbb{K})$, where $\mathbb{K}$ is a global field of equal characteristic [Dri]. In the course of this work, he remarks that the *cuspidal* characters of $G = \mathrm{SL}_2(\mathbb{F}_q)$ may be found in the first ℓ-adic cohomology group (here, ℓ is a prime number different from p) of the curve $\mathbf{Y}$ with equation $xy^q - yx^q = 1$, on which G acts naturally by linear changes of coordinates (we call $\mathbf{Y}$ the *Drinfeld curve*). This example inspired Deligne and Lusztig (see their comments in [DeLu, page 117, lines 22-24]) who then, in their fundamental article [DeLu], established the basis of what has come to be known as *Deligne-Lusztig theory*.

A large part of this book is concerned with unravelling Drinfeld's example. Our principal is to rather shamelessly make use of the fundamental results of ℓ-adic cohomology (for which we provide an overview tailored to our needs in Appendix A) to construct representations of G in characteristic 0 or ℓ. In order to efficiently use this machinery, we conduct a precise study of the geometric properties of the action of G on the Drinfeld curve $\mathbf{Y}$, with particular attention being paid to the construction of quotients by various finite groups.

Having completed this study we do not limit ourselves to character theory. Indeed, a large part of this book is dedicated to the study of modular representation theory, most notably via the study of Broué's abelian defect group conjecture [Bro]. This conjecture predicts the existence of an equivalence of derived categories when the defect group is abelian. For the representations of the group G in characteristic $\ell \notin \{2, p\}$, the defect group is cyclic, and such an equivalence can be obtained by entirely algebraic methods [Ric1], [Lin], [Rou2]. However, in order to stay true to the spirit of this book, we show that it is possible, when ℓ is odd and divides $q+1$, to realise this equivalence of derived categories using the complex of ℓ-adic cohomology of the Drinfeld curve (this result is due to Rouquier [Rou1]).

For completeness we devote a chapter to the study of representations in *equal*, or *natural*, characteristic. Here the Drinfeld curve ceases to be useful to us. We give an algebraic construction of the simple modules by restriction of rational representations of the group $\mathbf{G} = \mathrm{SL}_2(\overline{\mathbb{F}}_q)$, as may be done for an arbitrary finite reductive group. Moreover, in this case the Sylow p-subgroup is abelian, and it was shown by Okuyama [Oku1], [Oku2] (for the principal block) and Yoshii [Yo] (for the nonprincipal block with full defect) that Broué's conjecture holds. Unfortunately, the proof is too involved to be included in this book.

PREREQUISITES – The reader should have a basic knowledge of the representation theory of finite groups (as contained, for example, in [Ser] or [Isa]). In the appendix we recall the basics of *block theory*. He or she should also have a basic knowledge of algebraic geometry over an algebraically closed field (knowledge of the first chapter of [Har], for example, is more than sufficient). An overview of ℓ-adic cohomology is given in Appendix A, while Appendix C contains some basic facts about reflection groups (this appendix will only be used when we discuss some curiosities connected to the groups $\mathrm{SL}_2(\mathbb{F}_q)$ for $q \in \{3, 5, 7\}$).

We have also added a number of sections and subsections (marked with an asterisk) which contain illustrations, provided by G and $\mathbf{Y}$, of related but more geometric subjects (for example the Hurwitz formula, automorphisms of curves, Abyankhar's conjecture, invariants of reflections groups). These sections require a more sophisticated geometric background and are not necessary for an understanding of the main body of this book.

Lastly, for results concerning derived categories, we will not need more than is contained in the economical and efficient summary in Appendix A1 of the book of Cabanes and Enguehard [CaEn]. The sections and subsections requiring some knowledge of derived categories are also marked with an asterisk.

Besançon, France
April 2009

Cédric Bonnafé

Acknowledgements

This book probably would have never seen the light of day had it not been for the suggestion of Raphaël Rouquier, who also offered constant encouragement whilst the project was underway. He also patiently answered numerous questions concerning derived categories, and carefully read large passages of this book. For all of this I would like to thank him warmly.

I would also like to thank Marc Cabanes for numerous fruitful conversations on the subject of this book.

This English version of the book is the translation, by Geordie Williamson, of a French manuscript I had written. I want to thank him warmly for the quality of his job, as well as for the hundreds of interesting remarks, comments, critics, suggestions. It has been a great pleasure to work with him during the translation process.

Contents

List of Tables

General Notation

If $\mathscr{E}$ is a set, the cardinality of $\mathscr{E}$ (possibly infinite) will be denoted $|\mathscr{E}|$. If $\sim$ is an equivalence relation on $\mathscr{E}$, we will denote by $[\mathscr{E}/\sim]$ a set of representatives for the equivalence classes under $\sim$. When this notation is used the reader will have no difficulty verifying that the statement does not depend on the choice of representative. In an expression of the form $\sum_{x \in [\mathscr{E}/\sim]} f(x)$, the reader will also be able to verify that the element $f(x)$ depends only on the equivalence class of x.

If Γ is a group, we will denote by $\mathrm{Z}(\Gamma)$ its centre and $\mathrm{D}(\Gamma)$ its derived subgroup. If $\mathscr{E}$ is a subset of Γ, we denote by $\mathrm{N}_\Gamma(\mathscr{E})$ (respectively $\mathrm{C}_\Gamma(\mathscr{E})$) its normaliser (respectively centraliser). If $\gamma \in \Gamma$, we will denote by $\mathrm{Cl}_\Gamma(\gamma)$ its conjugacy class in Γ and we set ${}^\gamma\mathscr{E} = \gamma\mathscr{E}\gamma^{-1}$. Of course we have $|\Gamma| = |\mathrm{C}_\Gamma(\gamma)| \cdot |\mathrm{Cl}_\Gamma(\gamma)|$. The order (possibly infinite) of γ will be denoted $o(\gamma)$. If $\gamma' \in \Gamma$, we will denote by $[\gamma, \gamma']$ the commutator $\gamma\gamma'\gamma^{-1}\gamma'^{-1}$.

If A is a ring, we will denote by A^+ its underlying additive group and by $A^\times$ the (multiplicative) group of invertible elements in A. The centre of A will be denoted $\mathrm{Z}(A)$. We will denote by $A\text{-mod}$ the category of left A-modules of finite type. Given another ring B, we denote by $(A, B)\text{-bimod}$ the category of (A, B)-bimodules of finite type; that is, the category whose objects are simultaneously a left A-module of finite type and a right B-module of finite type satisfying $(a \cdot m) \cdot b = a \cdot (m \cdot b)$ for all $a \in A$, $b \in B$ and $m \in M$. Given a commutative ring R, an R-algebra A, a commutative R-algebra R' and an A-module M we will simplify the notation and denote by $R'M$ the extension of scalars $R' \otimes_R M$: it is an $R'A = R' \otimes_R A$-module. If in addition R is a field, we will denote by $\mathscr{K}_0(A)$ the Grothendieck group of the category of A-modules which are *finite dimensional* over R. If M is a left A-module which is finite dimensional over R, we will denote by $[M]$ (or $[M]_A$ if necessary) its class in $\mathscr{K}_0(A)$. We denote by $\mathrm{Irr}\, A$ a set of representatives for the isomorphism classes of simple (i.e. irreducible) A-modules which are finite dimensional over R; it follows that $\mathscr{K}_0(A)$ is the free $\mathbb{Z}$-module with basis $([S])_{S \in \mathrm{Irr}\, A}$.

We denote by $\mathrm{C}^b(A)$ the category of bounded complexes of A-modules, $\mathrm{K}^b(A)$ the bounded homotopy category of A-modules and by $\mathrm{D}^b(A)$ the derived category of A-modules (see [CaEn, §A1.2, A1.5 and A1.6]). After replacing A-modules by (A, B)-bimodules, we obtain the categories $\mathsf{C}^b(A, B)$, $\mathsf{K}^b(A, B)$ and $\mathsf{D}^b(A, B)$. If $\mathscr{C}$ and $\mathscr{C}'$ are two bounded complexes of A-modules (or of (A, B)-bimodules), we will write $\mathscr{C} \simeq_{\mathsf{C}} \mathscr{C}'$ (respectively $\mathscr{C} \simeq_{\mathsf{K}} \mathscr{C}'$, respectively $\mathscr{C} \simeq_{\mathsf{D}} \mathscr{C}'$) if $\mathscr{C}$ and $\mathscr{C}'$ are isomorphic in the category of complexes (respectively in the homotopy category, respectively in the derived category). We will denote by $\mathscr{C}[i]$ the complex shifted by i to the left [CaEn, §A1.2].

If $\mathscr{M}$ is a monoid and X is a subset of $\mathscr{M}$, we denote by $\langle X \rangle_{\mathrm{mon}}$ the submonoid of $\mathscr{M}$ generated by X. If R is a commutative ring, the monoid algebra of $\mathscr{M}$ over R will be denoted by $R\mathscr{M}$. If $\mathbb{K}$ is a field and M is a finite dimensional $\mathbb{K}\mathscr{M}$-module, we use $[M]_{\mathscr{M}}$ to denote the class of M in $\mathscr{K}_0(\mathbb{K}\mathscr{M})$ if there is no possible ambiguity as to the field $\mathbb{K}$.

If R is a ring and M is an R-module, we denote by $\mathsf{GL}_R(M)$ the group of its R-linear automorphisms. If n is a non-zero natural number, we will denote by $\mathsf{Mat}_n(R)$ the R-algebra of $n \times n$ square matrices with coefficients in R and set $\mathsf{GL}_n(R) = \mathsf{Mat}_n(R)^\times$. The identity matrix will be denoted I_n. Given $M \in \mathsf{Mat}_n(R)$, its transpose will be denoted ${}^tM$.

If R is commutative, the R-algebra $\mathsf{Mat}_n(R)$ as well as the group $\mathsf{GL}_n(R)$ will be naturally identified with $\mathsf{End}_R(R^n)$ and $\mathsf{GL}_R(R^n)$ respectively, using the canonical identification of R^n with the R-module of $n \times 1$ column vectors. We will denote by $\mathsf{SL}_n(R)$ the subgroup of $\mathsf{GL}_n(R)$ consisting of matrices of determinant 1. If $\mathbb{K}$ is a field, a matrix $g \in \mathsf{GL}_n(\mathbb{K})$ will be said to be *semisimple* if it is diagonalisable over an algebraic closure of $\mathbb{K}$. It will be said to be *unipotent* if $g - \mathsf{I}_n$ is nilpotent, that is, if $(g - \mathsf{I}_n)^n = 0$.

Part I
Preliminaries

The representation theory of $\mathrm{SL}_2(\mathbb{F}_q)$ will not commence until Chapter 3. Beforehand, in Chapters 1 and 2, we assemble the necessary preliminary facts needed for the study of representations. In Chapter 1 we are interested in the structure of the group $\mathrm{SL}_2(\mathbb{F}_q)$ (special subgroups, conjugacy classes, structure of centralisers, Sylow subgroups and their normalisers). Chapter 2 begins the study of the geometric properties of the Drinfeld curve: we essentially study certain interesting quotients as well as the action of the Frobenius endomorphism.

Chapter 1
Structure of $\mathrm{SL}_2(\mathbb{F}_q)$

Notation. Let p be an *odd* prime number, $\mathbb{F}$ an algebraic closure of the finite field $\mathbb{F}_p$ with p elements, q a power of p, $\mathbb{F}_q$ the subfield of $\mathbb{F}$ with q elements and G the group

$$G = \mathrm{SL}_2(\mathbb{F}_q) = \{\begin{pmatrix} x & z \\ y & t \end{pmatrix} \in \mathrm{GL}_2(\mathbb{F}_q) \mid xt - yz = 1\}.$$

If n is a non-zero natural number, we denote by μ_n the group of the n-th roots of unity in $\mathbb{F}^\times$:

$$\mu_n = \{\xi \in \mathbb{F}^\times \mid \xi^n = 1\}.$$

We will denote by $\mathrm{Tr}_2 : \mathbb{F}_{q^2} \to \mathbb{F}_q, \xi \mapsto \xi + \xi^q$ and $\mathrm{N}_2 : \mathbb{F}_{q^2}^\times \to \mathbb{F}_q^\times, \xi \mapsto \xi^{1+q}$ the trace and norm respectively.

1.1. Special Subgroups

Before commencing, recall that

(1.1.1) $$|G| = q(q-1)(q+1).$$

1.1.1. Bruhat Decomposition

Let B (respectively T, respectively U) denote the subgroup of G consisting of upper triangular (respectively diagonal, respectively unipotent upper triangular) matrices. Then

$$B = T \ltimes U.$$

C. Bonnafé, *Representations of* $\mathrm{SL}_2(\mathbb{F}_q)$, Algebra and Applications 13,
DOI 10.1007/978-0-85729-157-8_1,

We have isomorphisms

$$\mathbf{d}\colon \mathbb{F}_q^\times \xrightarrow{\sim} T, \quad a \mapsto \mathrm{diag}(a, a^{-1}) = \begin{pmatrix} a & 0 \\ 0 & a^{-1} \end{pmatrix}$$

and
$$\mathbf{u}\colon \mathbb{F}_q^+ \xrightarrow{\sim} U, \quad x \mapsto \begin{pmatrix} 1 & x \\ 0 & 1 \end{pmatrix}.$$

Proposition 1.1.2. *With the previous notation we have:*

(a) *The groups T and U are abelian and the group B is solvable.*
(b) *If $a \in \mathbb{F}_q^\times$ and $x \in \mathbb{F}_q$, then $\mathbf{d}(a)\mathbf{u}(x)\mathbf{d}(a)^{-1} = \mathbf{u}(a^2x)$.*

Now set
$$s = \begin{pmatrix} 0 & -1 \\ 1 & 0 \end{pmatrix}.$$

Then

(1.1.3) $$s^2 = -\mathrm{I}_2 \quad \text{and} \quad s\,\mathbf{d}(a)\,s^{-1} = \mathbf{d}(a^{-1}).$$

In particular, s normalizes T. The *Bruhat decomposition* of G is the partition of G into double cosets for the action of B:

(1.1.4) $$G = B \,\dot{\cup}\, BsB = B \,\dot{\cup}\, UsB.$$

Note also that

(1.1.5) $$B \cap {}^sB = T \quad \textit{and} \quad BsBsB = G.$$

Proof (of 1.1.4 and 1.1.5). Let $g = \begin{pmatrix} a & b \\ c & d \end{pmatrix} \in G \setminus B$. Then $c \neq 0$ and therefore $s^{-1}\mathbf{u}(-a/c)g \in B$. Hence $g \in UsB$. This proves 1.1.4.

The first equality of 1.1.5 is immediate. To establish the second equality let $X = BsBsB$. Then X is stable by left and right multiplication by B and $s^2 = -\mathrm{I}_2 \in X$. It follows that $X = B$ or $X = G$ by 1.1.4. But $s\mathbf{u}(1)s \in X \setminus B$ and so $X = G$. □

In what follows we denote by N the group $\langle T, s\rangle$. The exact sequence

(1.1.6) $$1 \longrightarrow T \longrightarrow N \longrightarrow \mathbb{Z}/2\mathbb{Z} \longrightarrow 1$$

is not split. In fact, all elements $g \in N \setminus T$ satisfy $g^2 = -\mathrm{I}_2$. We will see later that, if $q > 3$, then N is the normaliser of T (Corollary 1.3.2). It follows easily from the Bruhat decomposition that

(1.1.7) $$G = \dot{\bigcup_{n \in N}} UnU.$$

1.1.2. The Non-Split Torus

The choice of a basis of the $\mathbb{F}_q$-vector space $\mathbb{F}_{q^2}$ induces an isomorphism

$$\mathbf{d}' \colon \mathrm{GL}_{\mathbb{F}_q}(\mathbb{F}_{q^2}) \xrightarrow{\sim} \mathrm{GL}_2(\mathbb{F}_q).$$

The group $\mathbb{F}_{q^2}^\times$ acts on the $\mathbb{F}_q$-vector space $\mathbb{F}_{q^2}$ (which is of dimension 2) by multiplication. We may therefore view it as a subgroup of $\mathrm{GL}_{\mathbb{F}_q}(\mathbb{F}_{q^2})$. It is then easy to verify that, if $\xi \in \mathbb{F}_{q^2}^\times$, then

(1.1.8) $$\mathrm{Tr}\,\mathbf{d}'(\xi) = \xi^q + \xi = \mathrm{Tr}_2(\xi) \quad \text{and} \quad \det \mathbf{d}'(\xi) = \xi^{1+q} = \mathrm{N}_2(\xi).$$

(see Exercise 1.4 for an explicit construction of $\mathbf{d}'$ and a proof). Therefore the image T' of μ_{q+1} under $\mathbf{d}'$ is contained in G.

The subgroup T (respectively T') of G will be called the *split torus* (respectively the *non-split torus*, or *anisotropic torus*) of G. We have isomorphisms

(1.1.9) $$\mathbf{d} \colon \mu_{q-1} = \mathbb{F}_q^\times \xrightarrow{\sim} T \quad \text{and} \quad \mathbf{d}' \colon \mu_{q+1} \xrightarrow{\sim} T'.$$

The Frobenius automorphism $F \colon \mathbb{F}_{q^2} \to \mathbb{F}_{q^2}$, $\xi \mapsto \xi^q$ is $\mathbb{F}_q$-linear, and is therefore an element of $\mathrm{GL}_{\mathbb{F}_q}(\mathbb{F}_{q^2})$. Set $\tilde{s}' = \mathbf{d}'(F)$.We have

(1.1.10) $$\tilde{s}'^2 = \mathrm{I}_2 \quad \text{and} \quad \tilde{s}'\mathbf{d}'(\xi)\tilde{s}'^{-1} = \mathbf{d}'(\xi^q).$$

On the other hand,

(1.1.11) $$\det \tilde{s}' = -1.$$

In fact, $\tilde{s}'$ is of order 2 but does not commute with all the elements of $\mathrm{GL}_2(\mathbb{F}_q)$ by 1.1.10 and is therefore similar to the matrix $\mathrm{diag}(1,-1)$.

Fix an element ξ_0 of $\mathbb{F}_{q^2}^\times$ such that $\mathrm{N}_2(\xi_0) = -1$ (the norm $\mathrm{N}_2 \colon \mathbb{F}_{q^2}^\times \to \mathbb{F}_q^\times$ is surjective by Exercise 1.4(a)) and set

$$s' = \mathbf{d}'(\xi_0)\tilde{s}'.$$

Then $s' \in G$ and, if $\xi \in \mu_{q+1}$,

(1.1.12) $$s'^2 = -\mathrm{I}_2 \quad \text{and} \quad s'\,\mathbf{d}'(\xi)\,s'^{-1} = \mathbf{d}'(\xi^q) = \mathbf{d}'(\xi)^{-1}$$

In particular, s' normalizes T'. In what follows, the group $\langle T', s' \rangle$ will be denoted N'. The exact sequence

(1.1.13) $$1 \longrightarrow T' \longrightarrow N' \longrightarrow \mathbb{Z}/2\mathbb{Z} \longrightarrow 1$$

is not split. In fact, all the elements $g \in N' \setminus T'$ satisfy $g^2 = -\mathrm{I}_2$. We will see later that N' is the normaliser of T' (Corollary 1.3.2).

1.2. Distinguished Subgroups

It is easy to verify that

$$\text{(1.2.1)} \qquad \mathrm{Z}(G) = \{\mathrm{I}_2, -\mathrm{I}_2\}.$$

To simplify notation, we set $Z = \mathrm{Z}(G)$. We will show that, if $q > 3$, then G/Z is simple. Beforehand we will need the following two lemmas:

Lemma 1.2.2. *The group G is generated by U and sUs^{-1}. In particular, $G = \langle U, s \rangle$.*

Proof. Set $H = \langle U, sUs^{-1} \rangle$. By 1.1.4, it is enough to show that $s \in H$ and $T \subseteq H$. Two simple calculations show that

$$s = \mathbf{u}(-1) \cdot (s\mathbf{u}(1)s^{-1}) \cdot \mathbf{u}(-1) \in H$$

and, if $a \in \mathbb{F}_q^{\times}$, then

$$\mathbf{d}(a) = \mathbf{u}(a) \cdot (s\mathbf{u}(-a^{-1})s^{-1}) \cdot \mathbf{u}(a)s \in H,$$

which finishes the proof. □

Lemma 1.2.3. $\bigcap_{g \in G} gBg^{-1} = Z.$

Proof. Set $H = \bigcap_{g \in G} gBg^{-1}$. It is clear that $Z \subseteq H$. On the other hand, note that $B \cap {}^sB = T$. If we set $u = \mathbf{u}(1)$, then a calculation shows immediately that ${}^uT \cap T = Z$. The result then follows. □

Theorem 1.2.4. *We have:*

(a) *If $q > 3$, then $\mathrm{D}(G) = G$ and Z is the only non-trivial normal subgroup of G. In particular, G/Z is simple.*
(b) *If $q = 3$, the non-trivial normal subgroups of G are Z and N'. Moreover, $G = N' \rtimes U$ and $\mathrm{D}(G) = N'$.*

Proof. (a) Suppose that $q > 3$. We begin by showing that $G = \mathrm{D}(G)$. As $q > 3$, there exists $a \in \mathbb{F}_q^{\times}$ such that $a^2 \neq 1$. Now, if $x \in \mathbb{F}_q^{+}$, it follows from Proposition 1.1.2(b) that $[\mathbf{d}(a), \mathbf{u}(x)] = \mathbf{u}((a^2 - 1)x)$. This shows that $U \subseteq \mathrm{D}(G)$ and therefore that $\mathrm{D}(G) = G$ by Lemma 1.2.2.

Let H be a non-trivial normal subgroup of G. If H is contained in B, then H is contained in the intersection of the conjugates of B, that is $H \subseteq Z$ by Lemma 1.2.3. If H is not contained in B, then set $G' = HB$. It is a subgroup of G which strictly contains B, and so $G' = G$ thanks to the Bruhat decomposition 1.1.4. It follows that $G/H \simeq B/(B \cap H)$ and, as $\mathrm{D}(G) = G$, we obtain that $B/(B \cap H)$ is equal to its derived group. As B is solvable, we must have

$B \cap H = B$, and therefore H contains B and all of its conjugates. Therefore $G = HB = H$ by Lemma 1.2.2.

(b) is straightforward and will be shown in Chapter 11 (see Proposition 11.2.3), which treats particular cases related to small values of q. □

1.3. Conjugacy Classes

1.3.1. Centralisers

The following proposition describes the centralisers of certain elements of our group G. As we will see in the following section, the list of elements contained in the proposition is exhaustive up to conjugacy.

Proposition 1.3.1. *Let $g \in G$. Then:*

(a) *If $g \in \{\mathsf{I}_2, -\mathsf{I}_2\}$, then $\mathsf{C}_G(g) = G$.*
(b) *If $g \in U \setminus \{\mathsf{I}_2\}$, then $\mathsf{C}_G(g) = \{\mathsf{I}_2, -\mathsf{I}_2\} \times U = ZU$.*
(c) *If $g = \mathbf{d}(a)$ with $a \in \mu_{q-1} \setminus \{1, -1\}$, then $\mathsf{C}_G(g) = T$.*
(d) *If $g = \mathbf{d}'(\xi)$ with $\xi \in \mu_{q+1} \setminus \{1, -1\}$, then $\mathsf{C}_G(g) = T'$.*

Proof. (a) is immediate, and (b) and (c) follow via elementary calculation. Let us now show (d). Using the isomorphism $\mathbf{d}' \colon \mathsf{GL}_{\mathbb{F}_q}(\mathbb{F}_{q^2}) \xrightarrow{\sim} \mathsf{GL}_2(\mathbb{F}_q)$, it is enough to show

$$\mathsf{C}_{\mathsf{GL}_{\mathbb{F}_q}(\mathbb{F}_{q^2})}(\xi) = \mathbb{F}_{q^2}^\times.$$

Firstly, it is clear that $\mathbb{F}_{q^2}^\times$ is contained in the centraliser of ξ. On the other hand, fix $g \in \mathsf{GL}_{\mathbb{F}_q}(\mathbb{F}_{q^2})$ such that $g\xi = \xi g$. Set $\xi_0 = g(1) \in \mathbb{F}_{q^2}^\times$. We then have, for all a and b in $\mathbb{F}_q$,

$$g(a + b\xi) = g(a) + bg(\xi) = ag(1) + b\xi g(1) = \xi_0(a + b\xi)$$

as g and ξ commute. But, as $\xi \notin \{1, -1\}$, we have $\xi^q = \xi^{-1} \neq \xi$ and therefore $\mathbb{F}_{q^2} = \mathbb{F}_q \oplus \mathbb{F}_q\xi$. It follows that $g = \xi_0 \in \mathbb{F}_{q^2}^\times$. □

Corollary 1.3.2. *We have:*

(a) *If $q > 3$, then $\mathsf{C}_G(T) = T$ and $\mathsf{N}_G(T) = N$. If $q = 3$, then $T = \{\mathsf{I}_2, -\mathsf{I}_2\} = Z$ and $\mathsf{N}_G(T) = \mathsf{C}_G(T) = G$.*
(b) *$\mathsf{C}_G(T') = T'$ and $\mathsf{N}_G(T') = N'$.*

Proof. (a) The case where $q = 3$ is immediate. Suppose therefore that $q > 3$. It is clear that $T \subseteq \mathsf{C}_G(T)$. Now choose $g \in \mathsf{C}_G(T)$. As $q > 3$, there exists an element $a \in \mathbb{F}_q^\times$ not equal to 1 or -1. As a consequence, g commutes with $\mathbf{d}(a) \in T$ and therefore $g \in T$ by Proposition 1.3.1(c).

For the second equality, note first that $N \subseteq \mathrm{N}_G(T)$. On the other hand, let $g \in \mathrm{N}_G(T)$. Then there exists $b \in \mathbb{F}_q^\times$ such that $g\,\mathbf{d}(a)\,g^{-1} = \mathbf{d}(b)$. This is only possible if $a = b$ or $a = b^{-1}$. If $a = b$, then g commutes with $\mathbf{d}(a)$ and therefore belongs to T by Proposition 1.3.1(c). If $a = b^{-1}$, then sg commutes with $\mathbf{d}(a)$ and therefore belongs to T, again by Proposition 1.3.1(c). The result follows.

(b) is shown in the same way after remarking that μ_{q+1} must contain an element different from 1 and -1. □

1.3.2. *Parametrisation*

We denote by $\equiv$ the relation on $\mathbb{F}^\times$ defined by $x \equiv y$ if $y \in \{x, x^{-1}\}$. The equivalence classes of 1 and -1 contain a unique element, and all other classes contain two elements. We also fix an element $z_0 \in \mathbb{F}_q$ which is not a square in $\mathbb{F}_q$ (which is possible as q is odd). Set

$$u_+ = \begin{pmatrix} 1 & 1 \\ 0 & 1 \end{pmatrix} \quad \text{and} \quad u_- = \begin{pmatrix} 1 & z_0 \\ 0 & 1 \end{pmatrix}.$$

Theorem 1.3.3. *The group G consists of $q+4$ conjugacy classes. A set of representatives is given by*

$$\{\mathrm{I}_2, -\mathrm{I}_2\} \cup \{u_+, u_-, -u_+, -u_-\} \cup \{\mathbf{d}(a) \mid a \in [(\mu_{q-1} \setminus \{1,-1\})/\equiv]\}$$
$$\cup \{\mathbf{d}'(\xi) \mid \xi \in [(\mu_{q+1} \setminus \{1,-1\})/\equiv]\}.$$

Proof. Denote by $\mathscr{E}$ the set

$$\{\mathrm{I}_2, -\mathrm{I}_2\} \cup \{u_+, u_-, -u_+, -u_-\} \cup \{\mathbf{d}(a) \mid a \in [(\mu_{q-1} \setminus \{1,-1\})/\equiv]\}$$
$$\cup \{\mathbf{d}'(\xi) \mid \xi \in [(\mu_{q+1} \setminus \{1,-1\})/\equiv]\}.$$

Let g and g' be two distinct elements of $\mathscr{E}$. We will show that they are not conjugate in G. In fact, in most cases they are not conjugate in $\mathrm{GL}_2(\mathbb{F}_q)$ (which can be deduced by comparing the eigenvalues over $\mathbb{F}$). The only delicate case is to show that u_+ and u_- (respectively $-u_+$ and $-u_-$) are not conjugate in G (even though they are in $\mathrm{GL}_2(\mathbb{F}_q)$).

We will only prove this for u_+ and u_-, the other case follows in the same manner. Let us therefore suppose that there exists $h \in G$ such that $hu_+h^{-1} = u_-$. As $\mathrm{Ker}(u_+ - \mathrm{I}_2) = \mathrm{Ker}(u_- - \mathrm{I}_2) = \mathbb{F}_q \begin{pmatrix} 1 \\ 0 \end{pmatrix}$, h stabilises the line $\mathbb{F}_q \begin{pmatrix} 1 \\ 0 \end{pmatrix}$. It follows that $h \in B$. If we write $h = \begin{pmatrix} a & b \\ 0 & a^{-1} \end{pmatrix}$ with $a \in \mathbb{F}_q^\times$ and $b \in \mathbb{F}_q$ then a straightforward calculation shows that $z_0 = a^2$, which is impossible.

It remains to show that every element of G is conjugate, *in G*, to an element of $\mathscr{E}$. For this, one could prove it using linear algebra. We have chosen to prove it by a counting argument: it is enough to show that

$$\sum_{g\in\mathscr{E}} |\mathrm{Cl}_G(g)| = |G|.$$

Using Proposition 1.3.1, we obtain

$$\begin{aligned}\sum_{g\in\mathscr{E}} |\mathrm{Cl}_G(g)| &= \sum_{g\in\mathscr{E}} \frac{|G|}{|C_G(g)|}\\ &= 2+4\times\frac{(q-1)(q+1)}{2}+\frac{q-3}{2}\times q(q+1)+\frac{q-1}{2}\times q(q-1)\\ &= |G|,\end{aligned}$$

as expected. □

The results of this section are summarised in Table 1.1.

Table 1.1 Conjugacy classes

Representative	$\pm I_2$	$\mathbf{d}(a)$	$\mathbf{d}'(\xi)$	$\begin{pmatrix}\varepsilon & a\\ 0 & \varepsilon\end{pmatrix}$
		$a\in\mu_{q-1}\setminus\{\pm1\}$	$\xi\in\mu_{q+1}\setminus\{\pm1\}$	$\varepsilon\in\{\pm1\},\ a\in\mathbb{F}_q^\times$
Number of classes	2	$\frac{q-3}{2}$	$\frac{q-1}{2}$	4
Order	$o(\pm1)$	$o(a)$	$o(\xi)$	$p\cdot o(\varepsilon)$
Cardinality	1	$q(q+1)$	$q(q-1)$	$\frac{q^2-1}{2}$
Centraliser	G	T	T'	ZU

1.4. Sylow Subgroups

In the course of the description of Sylow subgroups of G, we will see that some of the special subgroups introduced in Section 1.1 (B, U, T, N, T', N'...) will occur as normalisers or centralizers of Sylow subgroups. The situation is somewhat more complicated for Sylow 2-subgroups.

1.4.1. *Sylow p-Subgroups*

The following proposition is immediate.

Proposition 1.4.1. *U is a Sylow p-subgroup of G. Moreover,*

$$C_G(U) = ZU \quad \textit{and} \quad N_G(U) = B.$$

1.4.2. *Other Sylow Subgroups*

Fix a prime number ℓ not equal to p and dividing the order of G. We denote by S_ℓ (respectively S'_ℓ) the ℓ-Sylow subgroup of T (respectively T'). Note that

(1.4.2) $$S_\ell \textit{ and } S'_\ell \textit{ are cyclic.}$$

By 1.1.1, ℓ divides $(q-1)(q+1)$. As $\gcd(q-1, q+1) = 2$ (recall that q is odd), we have the following.

Theorem 1.4.3. *Let ℓ be a prime number different from p which divides the order of G.*

(a) *If ℓ odd and divides $q-1$, then S_ℓ is a Sylow ℓ-subgroup of G. Moreover*

$$C_G(S_\ell) = T \quad \textit{and} \quad N_G(S_\ell) = N.$$

(b) *If ℓ is odd and divides $q+1$, then S'_ℓ is a Sylow ℓ-subgroup of G. Moreover*

$$C_G(S'_\ell) = T' \quad \textit{and} \quad N_G(S'_\ell) = N'.$$

(c) *If $q \equiv 1 \mod 4$ (respectively $q \equiv 3 \mod 4$), then $\langle S_2, s\rangle$ (respectively $\langle S'_2, s'\rangle$) is a Sylow 2-subgroup of G.*

(d) *Let S be a Sylow 2-subgroup of G. Then*

$$C_G(S) = Z \quad \textit{and} \quad |N_G(S)/S| = \begin{cases} 1 & \textit{if } q \equiv \pm 1 \mod 8, \\ 3 & \textit{if } q \equiv \pm 3 \mod 8. \end{cases}$$

Proof. (a) If ℓ is odd and divides $q-1$, then ℓ divides neither q nor $q+1$. This shows that S_ℓ is a Sylow ℓ-subgroup of G. Let $g \in S_\ell \setminus \{I_2\}$. Proposition 1.3.1(c) tells us that $C_G(g) = T$, which allows us to conclude easily that $C_G(S_\ell) = T$. For the second equality, note that $N \subseteq N_G(S_\ell)$. On the other hand, $N_G(S_\ell)$ normalizes $C_G(S_\ell) = T$, and is therefore contained in N (as ℓ is odd and divides $q-1$, we have $q > 3$ and we can therefore apply Corollary 1.3.2).

(b) is shown in the same way as (a).

(c) It is enough to remark that, if $q \equiv 1 \mod 4$ (respectively $q \equiv 3 \mod 4$), then 4 does not divide $q+1$ (respectively $q-1$). This allows us to determine the 2-valuation of $|G|$.

(d) First suppose that $q \equiv 3 \mod 4$. Set $S = \langle S_2', s' \rangle$. By (c), S is a Sylow 2-subgroup of G. Now there exists $g \in S_2' \setminus \{\mathrm{I}_2, -\mathrm{I}_2\}$ and therefore $\mathrm{C}_G(g) = T'$ by Proposition 1.3.1(c). In particular, $\mathrm{C}_G(S) \subseteq T'$. But, by 1.1.12, the only elements of T' which commute with s' are I_2 and $-\mathrm{I}_2$. The case where $q \equiv 1 \mod 4$ is treated similarly.

It remains to calculate the normaliser of S.

• If $q \equiv 1 \mod 8$, we may suppose, by (c), that $S = \langle S_2, s \rangle$. Then S_2 contains an element of order 8, which we denote by t. Now t and t^{-1} are the only elements of order 8 in S. Therefore, if $g \in \mathrm{N}_G(S)$, then $gtg^{-1} \in \{t, t^{-1}\}$, which implies that $g \in N$. Remember that we hope to show that $g \in S$. After multiplying by s, we may suppose that $g \in T$. But then $gsg^{-1} \in S$ and hence, as $sgs^{-1} = g^{-1}$, we have $gsg^{-1}s^{-1} = g^2 \in S$. Therefore $g^2 \in S_2$, which forces $g \in S_2$. It follows that $\mathrm{N}_G(S) = S$.

• If $q \equiv -1 \mod 8$, the result is shown in the same way.

• If $q \equiv \pm 3 \mod 8$, then $|S| = 8$ and, as $-\mathrm{I}_2$ is the only element of order 2 in G, we can deduce that S is the quaternionic group of order 8. It is well known that the group of outer automorphisms of S is isomorphic to $\mathfrak{S}_3$ (via the action on the three conjugacy classes of elements of order 4), therefore $\mathrm{N}_G(S)/S\mathrm{C}_G(S) = \mathrm{N}_G(S)/S$ is of order dividing 6. As it is, moreover, of odd order, it can only be equal to 1 or 3. In order to show that it is of order 3, we must construct an element of order 3 in the normaliser of S. We treat only the case where $q \equiv 5 \mod 8$, whereas the case where $q \equiv 3 \mod 8$ is relatively similar and left as an exercise (see Exercise 1.4).

So assume that $q \equiv 5 \mod 8$. We may suppose that $S = \langle S_2, s \rangle$. Denote by i an element of $\mathbb{F}_q$ for which $i^2 = -1$ (which exists as $q \equiv 1 \mod 4$). Set $\alpha = \dfrac{1+i}{2}$ and

$$g = \begin{pmatrix} -\alpha & \alpha \\ i\alpha & i\alpha \end{pmatrix}.$$

It is simple to verify that g belongs to G, normalizes S and satisfies $g^3 = \mathrm{I}_2$. The proof of the theorem is finished. □

Exercises

1.1. Let k be the subfield of $\mathbb{F}_q$ generated by $(\mathrm{Tr}_2(\xi))_{\xi \in \mu_{q+1}}$. Show that $k = \mathbb{F}_q$. (***Hint:*** Set $q' = |k|$ and show that, if $\xi \in \mu_{q+1}$, then $1 + \xi^2 + \xi^{q'} + \xi^{q'+1} = 0$.)

1.2. Show that the map $U \times B \to BsB$, $(u, b) \mapsto usb$ is bijective.

1.3. Show that $\mathrm{N}_G(B) = B$.

1.4. The purpose of this exercise is to choose a basis over $\mathbb{F}_q$ of $\mathbb{F}_{q^2}$ which allows an easy and explicit construction of the matrices $\mathbf{d}'(\xi)$ for $\xi \in \mathbb{F}_{q^2}^\times$ and, in particular, to deduce a proof of 1.1.8.

(a) Show that Tr_2 and N_2 are surjective.
(b) For the rest of this exercise fix an element $z \in \mathbb{F}_{q^2}$ such that $z \neq 0$ and $\mathrm{Tr}_2(z) = 0$ (i.e. $z^q + z = 0$). Show that $z \notin \mathbb{F}_q$ and conclude that $(1, z)$ is a $\mathbb{F}_q$-basis of $\mathbb{F}_{q^2}$. We construct the isomorphism $\mathbf{d}' \colon \mathrm{GL}_{\mathbb{F}_q}(\mathbb{F}_{q^2}) \xrightarrow{\sim} \mathrm{GL}_2(\mathbb{F}_q)$ using the basis $(1, z)$.
(c) Show that then $\tilde{s}' = \mathrm{diag}(1, -1)$.
(d) Let $\xi \in \mathbb{F}_{q^2}^\times$. Show that

$$\mathbf{d}'(\xi) = \begin{pmatrix} \dfrac{\xi^q + \xi}{2} & \dfrac{z(\xi - \xi^q)}{2} \\[2ex] \dfrac{\xi - \xi^q}{2z} & \dfrac{\xi^q + \xi}{2} \end{pmatrix}.$$

(e) Show that $\mathrm{Tr}\,\mathbf{d}'(\xi) = \mathrm{Tr}_2(\xi)$, $\det \mathbf{d}'(\xi) = \mathrm{N}_2(\xi)$ and that the characteristic polynomial of $\mathbf{d}'(\xi)$ is $(X - \xi)(X - \xi^q)$, where X is an indeterminate.
(f) Show that $\mathbf{d}'(\xi)$ is conjugate, *in* $\mathrm{GL}_2(\mathbb{F}_{q^2})$, to $\mathrm{diag}(\xi, \xi^q)$.
(g) Suppose that $q \equiv 3 \mod 8$. Set $S = \langle S_2', s' \rangle$ and denote by i an element of μ_{q+1} such that $i^2 = -1$.
   (g1) Show that $S_2' = \langle \mathbf{d}'(i) \rangle$ and calculate $\mathbf{d}'(i)$.
   (g2) Find a matrix $g \in G$ such that $gs'g^{-1} = \mathbf{d}'(i)$, $g\mathbf{d}'(i)g^{-1} = s'\mathbf{d}'(i)$ and show that g normalizes S and is of order 3 or 6.
   (g3) Conclude that $|\mathrm{N}_G(S)/S| = 3$ (which completes the proof of Theorem 1.4.3).

1.5. Denote by σ (respectively σ') the automorphism of T (respectively T') which sends an element to its inverse. Set $M = \langle \sigma \rangle \ltimes T$ and $M' = \langle \sigma' \rangle \ltimes T'$. Let R be a commutative ring in which 2 is invertible. Show that the group algebras RN and RM (respectively RN' and RM') are isomorphic.

1.6. Let H be a subgroup of T (respectively T') which is not contained in Z. Show that $\mathrm{C}_G(H) = T$ and $\mathrm{N}_G(H) = N$ (respectively $\mathrm{C}_G(H) = T'$ and $\mathrm{N}_G(H) = N'$).

1.7. Let n be a non-zero natural number and $g \in \mathrm{GL}_n(\mathbb{F})$. Show that g is semi-simple (respectively unipotent) if and only if its order is prime to p (respectively a power of p). **Remark:** Note that g is of finite order as $\mathbb{F} = \cup_{r \geqslant 1} \mathbb{F}_{p^r}$.

1.8. Show that u_+ and u_- are conjugate in $\mathrm{SL}_2(\mathbb{F}_{q^2})$ (although they are not in $G = \mathrm{SL}_2(\mathbb{F}_q)$). Show that they are also conjugate in $\mathrm{GL}_2(\mathbb{F}_q)$.

Show that two semi-simple elements of G are conjugate in G if and only if they are conjugate in $\mathrm{GL}_2(\mathbb{F}_q)$.

1.9. Show that the number of unipotent elements of G is q^2.

1.10. Show that the group G has q conjugacy classes of semi-simple elements.

1.11. Let r be a prime number and let P and Q be two distinct Sylow r-subgroup of G. Show that:

(a) If $r > 2$, then $P \cap Q = \{\mathrm{I}_2\}$.
(b) If $r = 2$, then $P \cap Q = \{\mathrm{I}_2, -\mathrm{I}_2\}$.

1.12. Let P be a Sylow 2-subgroup of G. Show that $P/\mathrm{D}(P) \simeq \mathbb{Z}/2\mathbb{Z} \times \mathbb{Z}/2\mathbb{Z}$.

1.13. Let $\mathbb{K}$ be a commutative field of characteristic different from 2 and (e_1, e_2) the canonical basis of $\mathbb{K}^2$. Consider $\det : \mathbb{K}^2 \times \mathbb{K}^2 \to \mathbb{K}$, the alternating bilinear (i.e. symplectic) form given by the determinant in the canonical basis. Let $J = \begin{pmatrix} 0 & 1 \\ -1 & 0 \end{pmatrix}$. Then J is the matrix of the symplectic form det in the canonical basis. Finally, denote by $\mathrm{Sp}_2(\mathbb{K})$ the automorphism group of $\mathbb{K}^2$ which stabilise the symplectic form det.

(a) Show that $\mathrm{Sp}_2(\mathbb{K}) = \mathrm{SL}_2(\mathbb{K})$.
(b) Show that $\mathrm{Sp}_2(\mathbb{K}) = \{g \in \mathrm{GL}_2(\mathbb{K}) \mid {}^t gJg = J\}$.
(c) Deduce that the automorphism of $\mathrm{SL}_2(\mathbb{K})$ defined by $g \mapsto {}^t g^{-1}$ is inner (and is induced by conjugation by J).
(d) Show that the automorphism of $\mathrm{GL}_2(\mathbb{K})$ given by $g \mapsto {}^t g^{-1}$ is not inner.

1.14*. In this exercise we calculate the group of outer automorphisms of $G = \mathrm{SL}_2(\mathbb{F}_q)$ and $\widetilde{G} = \mathrm{GL}_2(\mathbb{F}_q)$. Let $\phi : \mathbb{F}_q \to \mathbb{F}_q$, $x \mapsto x^p$ be the Frobenius automorphism. We will also denote by ϕ the automorphism of G or $\widetilde{G}$ induced by ϕ. Fix an element $a_0 \in \mathbb{F}_q^\times$ which is not a square and denote by σ the automorphism of G induced by conjugation by $\mathrm{diag}(a_0, 1) \in \widetilde{G}$.

(a) Show that σ and ϕ are non-inner automorphisms of G.

Denote by $\mathrm{Aut}(G)$ (respectively $\mathrm{Inn}(G)$, respectively $\mathrm{Out}(G)$) the group of automorphisms of G (respectively inner automorphisms of G, respectively outer automorphisms of G, i.e. $\mathrm{Out}(G) = \mathrm{Aut}(G)/\mathrm{Inn}(G)$). If $\gamma \in \mathrm{Aut}(G)$, we will denote by $\bar{\gamma}$ its image in $\mathrm{Out}(G)$. If $g \in G$, we set $\mathrm{inn}\, g : G \to G$, $h \mapsto ghg^{-1}$.

(b) Show that $\bar{\sigma}$ and $\bar{\phi}$ commute.
(c) Write $q = p^e$. Show that $\bar{\sigma}^2 = \bar{\phi}^e = \bar{\mathrm{Id}}_G$.
(d) Let γ be an automorphism of G. Show that there exists $g \in G$ such that $\gamma \circ \mathrm{inn}\, g$ stabilises B, U and T. (***Hint:*** Use the fact that U is a Sylow p-subgroup, that B is its normaliser and that all the Sylow p-subgroups of G are conjugate in G.)

(e) Deduce that $\mathrm{Out}(G) = \langle\bar{\sigma}\rangle \times \langle\bar{\phi}\rangle \simeq \mathbb{Z}/2\mathbb{Z} \times \mathbb{Z}/e\mathbb{Z}$.
Denote by $\tau\colon \widetilde{G} \to \widetilde{G}, g \mapsto {}^t g^{-1}$. Recall that τ is not an inner automorphism of $\widetilde{G}$ (see Exercise 1.13(d)).

(f) Show that $\mathrm{Out}(\widetilde{G}) = \langle\bar{\tau}\rangle \times \langle\bar{\phi}\rangle \simeq \mathbb{Z}/2\mathbb{Z} \times \mathbb{Z}/e\mathbb{Z}$.

Chapter 2
The Geometry of the Drinfeld Curve

Let $\mathbf{Y}$ be the *Drinfeld curve*

$$\mathbf{Y} = \{(x,y) \in \mathbf{A}^2(\mathbb{F}) \mid xy^q - yx^q = 1\}.$$

It is straightforward to verify that:

- G acts linearly on $\mathbf{A}^2(\mathbb{F})$ (via $g \cdot (x,y) = (ax+by, cx+dy)$ if $g = \begin{pmatrix} a & b \\ c & d \end{pmatrix} \in G$) and stabilises $\mathbf{Y}$;
- μ_{q+1} acts on $\mathbf{A}^2(\mathbb{F})$ by homotheties (via $\xi \cdot (x,y) = (\xi x, \xi y)$ if $\xi \in \mu_{q+1}$) and stabilises $\mathbf{Y}$;
- the Frobenius endomorphism $F\colon \mathbf{A}^2(\mathbb{F}) \to \mathbf{A}^2(\mathbb{F})$, $(x,y) \mapsto (x^q, y^q)$ stabilises $\mathbf{Y}$.

Moreover, if $g \in G$ and $\xi \in \mu_{q+1}$, then, as endomorphisms of $\mathbf{A}^2(\mathbb{F})$ (or $\mathbf{Y}$), we have

$$\begin{aligned} g \circ \xi &= \xi \circ g, \\ g \circ F &= F \circ g, \\ F \circ \xi &= \xi^{-1} \circ F. \end{aligned}$$

We can therefore form the monoid

$$G \times (\mu_{q+1} \rtimes \langle F \rangle_{\text{mon}})$$

which acts on $\mathbf{A}^2(\mathbb{F})$ and stabilises $\mathbf{Y}$.

The purpose of this chapter is to assemble the geometric properties of $\mathbf{Y}$ and the action of $G \times (\mu_{q+1} \rtimes \langle F \rangle_{\text{mon}})$ which allows us to calculate its ℓ-adic cohomology (as a module for the monoid $G \times (\mu_{q+1} \rtimes \langle F \rangle_{\text{mon}})$). A large part of this chapter is dedicated to the construction of quotients of $\mathbf{Y}$ by the actions of the finite groups G, U and μ_{q+1}.

C. Bonnafé, *Representations of* $\mathrm{SL}_2(\mathbb{F}_q)$, Algebra and Applications 13,
DOI 10.1007/978-0-85729-157-8_2, © Springer-Verlag London Limited 2011

2.1. Elementary Properties

The following proposition is (almost) immediate.

Proposition 2.1.1. *The curve* $\mathbf{Y}$ *is affine, smooth and irreducible.*

Proof. $\mathbf{Y}$ is affine because it is a closed subspace of the affine space $\mathbf{A}^2(\mathbb{F})$. It is irreducible because the polynomial $XY^q - YX^q - 1$ in $\mathbb{F}[X, Y]$ is irreducible (See Exercise 2.1). It is smooth because the differential of this polynomial is given by the 1×2 matrix $\begin{pmatrix} Y^q & -X^q \end{pmatrix}$, which is zero only at $(0,0) \notin \mathbf{Y}$. □

Proposition 2.1.2. *The group* G *acts freely on* $\mathbf{Y}$.

Proof. Let $g \in G$ and $(x,y) \in \mathbf{Y}$ be such that $g \cdot (x,y) = (x,y)$. It follows that 1 is an eigenvalue of g and, after conjugating g by an element of G, we may assume that there exists an $a \in \mathbb{F}_q$ such that

$$g = \begin{pmatrix} 1 & a \\ 0 & 1 \end{pmatrix}.$$

Then $x + ay = x$ and, as $y \neq 0$ (since $(x,y) \in \mathbf{Y}$), we conclude that $a = 0$. □

The next proposition is clear.

Proposition 2.1.3. *The group* μ_{q+1} *acts freely on* $\mathbf{A}^2(\mathbb{F}) \setminus \{(0,0)\}$ *and therefore also on* $\mathbf{Y}$.

Note however, that the group $G \times \mu_{q+1}$ does not act freely on $\mathbf{Y}$: the pair $(-\mathrm{I}_2, -1)$ acts as the identity. (Even the quotient $(G \times \mu_{q+1})/\langle(-\mathrm{I}_2, -1)\rangle$ does not act freely, see Exercise 2.3.)

2.2. Interesting Quotients

We will now describe the quotients of $\mathbf{Y}$ by the finite groups G, U and μ_{q+1}. In order to construct them we will use the following proposition, a proof of which can be found in [Bor, Proposition 6.6]. (Note that the proposition is far from optimal, but will be sufficient for our needs.)

Proposition 2.2.1. *Let* $\mathbf{V}$ *and* $\mathbf{W}$ *be two smooth and irreducible varieties,* $\varphi : \mathbf{V} \to \mathbf{W}$ *a morphism of varieties, and* Γ *a finite group acting on* $\mathbf{V}$. *Suppose that the following three properties are satisfied:*

(1) φ *is surjective;*
(2) $\varphi(v) = \varphi(v')$ *if and only if* v *and* v' *are in the same* Γ*-orbit;*
(3) *There exists* $v_0 \in \mathbf{V}$ *such that the differential of* φ *at* v_0 *is surjective.*

Then the morphism $\bar{\varphi} : \mathbf{V}/\Gamma \longrightarrow \mathbf{W}$ *induced by* φ *is an isomorphism of varieties.*

2.2.1. Quotient by G

The map

$$\begin{array}{rcl}\gamma\colon\quad \mathbf{Y} & \longrightarrow & \mathbf{A}^1(\mathbb{F}) \\ (x,y) & \longmapsto & xy^{q^2}-yx^{q^2}\end{array}$$

is a morphism of varieties. It is $\mu_{q+1} \rtimes \langle F \rangle_{\text{mon}}$-equivariant (for the action of μ_{q+1} on $\mathbf{A}^1(\mathbb{F})$ given by $\xi \cdot z = \xi^2 z$ and the action of F given by $z \mapsto z^q$). An elementary calculation shows that γ is constant on G-orbits. Even better, if we denote by $\bar{\gamma}\colon \mathbf{Y}/G \to \mathbf{A}^1(\mathbb{F})$ the morphism of varieties obtained by passing to the quotient, we have the following.

Theorem 2.2.2. *The morphism of varieties $\bar{\gamma}\colon \mathbf{Y}/G \to \mathbf{A}^1(\mathbb{F})$ is a $\mu_{q+1} \rtimes \langle F \rangle_{\text{mon}}$-equivariant isomorphism.*

Proof. The $\mu_{q+1} \rtimes \langle F \rangle_{\text{mon}}$-equivariance is evident. In order to show that $\bar{\gamma}$ is an isomorphism we must verify points (1), (2) and (3) of Proposition 2.2.1.

Choose $a \in \mathbb{F}$. To show (1) and (2), it is sufficient to show that $|\gamma^{-1}(a)| = |G|$ (as G acts freely on $\mathbf{Y}$ by Proposition 2.1.2). After changing variables $(z,t) = (x, y/x)$, we have a bijection $\gamma^{-1}(a) \xrightarrow{\sim} \mathscr{E}_a$, where

$$\mathscr{E}_a = \{(z,t) \in \mathbb{F}^\times \times \mathbb{F}^\times \mid t^q - t = \frac{1}{z^{q+1}} \text{ and } t^{q^2} - t = \frac{a}{z^{q^2+1}}\}.$$

As $t^{q^2} - t = (t^q - t)^q + (t^q - t)$, we obtain

$$\mathscr{E}_a = \{(z,t) \in \mathbb{F}^\times \times \mathbb{F}^\times \mid t^q - t = \frac{1}{z^{q+1}} \text{ and } \frac{1}{z^{q+1}} + \frac{1}{z^{q^2+q}} = \frac{a}{z^{q^2+1}}\}.$$

Or equivalently

$$\mathscr{E}_a = \{(z,t) \in \mathbb{F}^\times \times \mathbb{F}^\times \mid z^{q^2-1} - az^{q-1} + 1 = 0 \text{ and } t^q - t = \frac{1}{z^{q+1}}\}.$$

The polynomial $z^{q^2-1} - az^{q-1} + 1$ is coprime to its derivative, and therefore has $q^2 - 1$ distinct non-zero roots. For each of these roots, there are q non-zero solutions t to the equation $t^q - t = \frac{1}{z^{q+1}}$. Therefore

$$|\gamma^{-1}(a)| = |\mathscr{E}_a| = (q^2-1)q = |G|,$$

as expected.

We now turn to (3). Let $v = (x_0, y_0) \in \mathbf{Y}$. The tangent space $\mathscr{T}_v(\mathbf{Y})$ to $\mathbf{Y}$ at v has equation $y_0^q x - x_0^q y = 0$ and the differential $d_v\gamma\colon \mathscr{T}_v(\mathbf{Y}) \to \mathbb{F} = \mathscr{T}_{\gamma(v)}(\mathbf{A}^1(\mathbb{F}))$ is given by

$$d_v\gamma(x,y) = y_0^{q^2} x - x_0^{q^2} y.$$

Therefore, if $(x,y) \in \operatorname{Ker} d_v\gamma$, then

$$y_0^q x - x_0^q y = 0 \quad \text{and} \quad y_0^{q^2} x - x_0^{q^2} y = 0.$$

The determinant of this system is $-y_0^q x_0^{q^2} + x_0^q y_0^{q^2} = (x_0 y_0^q - y_0 x_0^q)^q = 1$, therefore $\operatorname{Ker} d_v\gamma = 0$. □

2.2.2. *Quotient by U*

The morphism

$$\begin{array}{rccc} \upsilon: & \mathbf{Y} & \longrightarrow & \mathbf{A}^1(\mathbb{F}) \setminus \{0\} \\ & (x,y) & \longmapsto & y \end{array}$$

is well-defined and is a morphism of varieties. It is $\mu_{q+1} \rtimes \langle F \rangle_{\mathrm{mon}}$-equivariant (for the action of μ_{q+1} on $\mathbf{A}^1(\mathbb{F}) \setminus \{0\}$ given by $\xi \cdot z = \xi z$ and the action of F given by $z \mapsto z^q$). An elementary calculation show that υ is constant on U-orbits. Even better, if we denote by $\bar{\upsilon}: \mathbf{Y}/U \to \mathbf{A}^1(\mathbb{F}) \setminus \{0\}$ the morphism of varieties induced by passing to the quotient, we have the following.

Theorem 2.2.3. *The morphism of varieties* $\bar{\upsilon}: \mathbf{Y}/U \to \mathbf{A}^1(\mathbb{F}) \setminus \{0\}$ *is a* $\mu_{q+1} \rtimes \langle F \rangle_{\mathrm{mon}}$*-equivariant isomorphism.*

Proof. The $\mu_{q+1} \rtimes \langle F \rangle_{\mathrm{mon}}$-equivariance is evident. To show that $\bar{\upsilon}$ is an isomorphism, we verify points (1), (2) and (3) of Proposition 2.2.1.

The surjectivity of υ is clear. We also have

$$\upsilon(x,y) = \upsilon(x',y') \Longleftrightarrow \exists\, u \in U,\ (x',y') = u \cdot (x,y).$$

Indeed, if $(x,y) \in \mathbf{Y}$ and $(x',y') \in \mathbf{Y}$ are such that $y = y'$, then

$$\left(\frac{x}{y}\right)^q - \frac{x}{y} = \left(\frac{x'}{y}\right)^q - \frac{x'}{y},$$

which shows that $\dfrac{x'-x}{y} \in \mathbb{F}_q$. Now, if we set $a = \dfrac{x'-x}{y}$, then

$$\begin{pmatrix} 1 & a \\ 0 & 1 \end{pmatrix} \cdot \begin{pmatrix} x \\ y \end{pmatrix} = \begin{pmatrix} x' \\ y' \end{pmatrix}.$$

This shows (2). Point (3) is immediate. □

2.2.3. *Quotient by* μ_{q+1}

The morphism

$$\begin{aligned} \pi: \quad \mathbf{Y} &\longrightarrow \mathbf{P}^1(\mathbb{F}) \setminus \mathbf{P}^1(\mathbb{F}_q) \\ (x,y) &\longmapsto \quad [x:y] \end{aligned}$$

is well-defined and is $G \times \langle F \rangle_{\mathrm{mon}}$-equivariant morphism of varieties (for the action of G induced by the natural action on $\mathbf{P}^1(\mathbb{F})$ and the action of F given by $[x;y] \mapsto [x^q;y^q]$). An elementary calculation show that π is constant on μ_{q+1}-orbits. Even better, if we denote by $\bar{\pi}: \mathbf{Y}/\mu_{q+1} \to \mathbf{P}^1(\mathbb{F}) \setminus \mathbf{P}^1(\mathbb{F}_q)$ the morphism of varieties induced by passage to the quotient, we have the following.

Theorem 2.2.4. *The morphism of varieties* $\bar{\pi}: \mathbf{Y}/\mu_{q+1} \to \mathbf{P}^1(\mathbb{F}) \setminus \mathbf{P}^1(\mathbb{F}_q)$ *is a* $G \times \langle F \rangle_{\mathrm{mon}}$*-equivariant isomorphism.*

Proof. The $G \times \langle F \rangle_{\mathrm{mon}}$-equivariance is evident. To show that $\bar{\pi}$ is an isomorphism, we should verify points (1), (2) and (3) of Proposition 2.2.1, which is straightforward. □

2.3. Fixed Points under certain Frobenius Endomorphisms

In order to get the most out of the Lefschetz fixed-point theorem (see Theorem A.2.7(a) in Appendix A) we will need the following two results. Firstly, note that, if $\xi \in \mu_{q+1}$, we have

(2.3.1) $$\mathbf{Y}^{\xi F} = \varnothing.$$

Indeed, $(\mathbf{Y}/\mu_{q+1})^F = \varnothing$ by Theorem 2.2.4. On the other hand, we have the following.

Theorem 2.3.2. *Let* $\xi \in \mu_{q+1}$. *Then*

$$|\mathbf{Y}^{\xi F^2}| = \begin{cases} 0 & \textit{if } \xi \neq -1, \\ q^3 - q & \textit{if } \xi = -1. \end{cases}$$

Proof. Let $(x,y) \in \mathbf{Y}^{\xi F^2}$. We then have

$$x = \xi x^{q^2}, \quad y = \xi y^{q^2} \quad \text{and} \quad xy^q - yx^q = 1.$$

As a consequence,

$$1 = (xy^q - yx^q)^q = x^q y^{q^2} - y^q x^{q^2} = \xi(x^q y - xy^q) = -\xi.$$

This shows that, if $\xi \neq -1$, then $\mathbf{Y}^{\xi F^2} = \varnothing$.

Therefore suppose that $\xi = -1$. We are looking for the number of solutions to the system

$$\begin{cases} x = -x^{q^2} & (1) \\ xy^q - yx^q = 1 & (2) \\ y = -y^{q^2} & (3) \end{cases}$$

However, if the pair (x, y) satisfies (1) and (2), then it also satisfies (3). Indeed, if (x, y) satisfies (1) and (2), then $x \neq 0$, $y^q = \dfrac{1 + yx^q}{x}$ and therefore

$$y^{q^2} = \left(\frac{1 + yx^q}{x}\right)^q = \frac{1 + y^q x^{q^2}}{x^q} = \frac{1 - xy^q}{x^q} = \frac{-yx^q}{x^q} = -y.$$

It follows that it is sufficient to find the number of solutions to the system given by equations (1) and (2). Now, x being non-zero, there are $q^2 - 1$ possibilities for x to be a solution of (1). As soon as we have fixed x, there are q solutions to equation (2) (viewed as an equation in y). Indeed, as an equation in y, $xy^q - yx^q - 1$ has derivative $-x^q \neq 0$, and so this polynomial does not admit multiple roots. This gives therefore $(q^2 - 1)q$ solutions to equations (1) and (2), and the theorem follows. □

REMARK – As G acts freely on $\mathbf{Y}$, the set $\mathbf{Y}^{-F^2}$ consists of a single G-orbit. □

2.4. Compactification

We will denote by $[x; y; z]$ homogeneous coordinates on the projective space $\mathbf{P}^2(\mathbb{F})$. We view $\mathbf{A}^2(\mathbb{F})$ as the open subset of $\mathbf{P}^2(\mathbb{F})$ defined by

$$\mathbf{A}^2(\mathbb{F}) \simeq \{[x; y; z] \in \mathbf{P}^2(\mathbb{F}) \mid z \neq 0\}.$$

We identify $\mathbf{P}^2(\mathbb{F}) \setminus \mathbf{A}^2(\mathbb{F})$ with $\mathbf{P}^1(\mathbb{F})$ (using the canonical isomorphism $[x; y] \mapsto [x; y; 0]$). The action of $G \times (\boldsymbol{\mu}_{q+1} \rtimes \langle F \rangle)$ on $\mathbf{A}^2(\mathbb{F})$ extends uniquely to $\mathbf{P}^2(\mathbb{F})$: if $g = \begin{pmatrix} a & b \\ c & d \end{pmatrix}$, $\xi \in \boldsymbol{\mu}_{q+1}$ and $[x; y; z] \in \mathbf{P}^2(\mathbb{F})$, then

$$g \cdot [x; y; z] = [ax + by; cx + dy; z],$$

$$\xi \cdot [x; y; z] = [\xi x; \xi y; z]$$

and
$$F[x; y; z] = [x^q; y^q; z^q].$$

Now let $\overline{\mathbf{Y}}$ be the projective curve defined by

$$\overline{\mathbf{Y}} = \{[x; y; z] \in \mathbf{P}^2(\mathbb{F}) \mid xy^q - yx^q = z^{q+1}\}.$$

The morphism

$$\begin{array}{rcl} \mathbf{Y} & \longrightarrow & \overline{\mathbf{Y}} \\ (x,y) & \longmapsto & [x;y;1] \end{array}$$

is an open immersion and allows us to identify $\mathbf{Y}$ with $\overline{\mathbf{Y}} \cap \mathbf{A}^2(\mathbb{F})$.

Proposition 2.4.1. *The closed subvariety $\overline{\mathbf{Y}}$ of $\mathbf{P}^2(\mathbb{F})$ is the closure of $\mathbf{Y}$ in $\mathbf{P}^2(\mathbb{F})$. It is smooth and stable under the action of of $G \times (\mu_{q+1} \rtimes \langle F \rangle)$. Moreover,*

$$\overline{\mathbf{Y}} \setminus \mathbf{Y} \simeq \mathbf{P}^1(\mathbb{F}_q),$$

with this isomorphism given by $[x;y] \in \mathbf{P}^1(\mathbb{F}_q) \mapsto [x;y;0]$.

Proof. The only point needing a little work is the smoothness. The points of $\mathbf{Y}$ are smooth by Proposition 2.1.1. As G acts transitively on $\overline{\mathbf{Y}} \setminus \mathbf{Y} = \mathbf{P}^1(\mathbb{F}_q) \simeq G/B$ (as a G-set), it is enough to show that $[1;0;0]$ is a smooth point of $\overline{\mathbf{Y}}$. For this, let us consider the open subvariety defined by $x \neq 0$. In this open set (again isomorphic to $\mathbf{A}^2(\mathbb{F})$, this time via the morphism $(y,z) \mapsto [1;y;z]$) $\overline{\mathbf{Y}}$ is defined by the equation $y - y^q - z^{q+1} = 0$ and the differential at $(0,0)$ of this polynomial is the 1×2 matrix

$$(1 \quad 0),$$

which is non-zero. □

We finish with a study of the quotient of $\overline{\mathbf{Y}}$ by μ_{q+1}. Consider the morphism

$$\begin{array}{rrcl} \pi_0 : & \overline{\mathbf{Y}} & \longrightarrow & \mathbf{P}^1(\mathbb{F}) \\ & [x;y;z] & \longmapsto & [x;y]. \end{array}$$

It is well-defined, G-equivariant, and surjective. Moreover, it is constant on μ_{q+1}-orbits and therefore induces, after passing to the quotient, a morphism of varieties $\bar{\pi}_0 : \overline{\mathbf{Y}}/\mu_{q+1} \to \mathbf{P}^1(\mathbb{F})$.

Theorem 2.4.2. *The morphism of varieties $\bar{\pi}_0 : \overline{\mathbf{Y}}/\mu_{q+1} \to \mathbf{P}^1(\mathbb{F})$ is a $G \times \langle F \rangle_{\mathrm{mon}}$-equivariant isomorphism.*

Proof. We omit the proof, as it follows the same arguments as those used in the proof of Theorem 2.2.4. □

2.5. Curiosities*

Independent of representation theory, the Drinfeld curve has interesting geometric properties which we discuss briefly here: it has a "large" automorphism group and gives a solution to a particular case of the *Abhyankar's Conjecture* [Abh] about unramified coverings of the affine line in positive characteristic.

2.5.1. *Hurwitz Formula, Automorphisms**

The group μ_{q+1} acts trivially on $\overline{\mathbf{Y}} \setminus \mathbf{Y} = \mathbf{P}^1(\mathbb{F}_q)$. Also, as μ_{q+1} is of order prime to p, the morphism π_0 is tamely ramified: it is only ramified at the points $a \in \mathbf{P}^1(\mathbb{F}_q)$ and ramification index at a is $e_a = q+1$. If we denote by $\mathbf{g}(\overline{\mathbf{Y}})$ the genus of $\overline{\mathbf{Y}}$, then

(2.5.1) $$\mathbf{g}(\overline{\mathbf{Y}}) = \frac{q(q-1)}{2}$$

as $\overline{\mathbf{Y}}$ is a smooth plane curve of degree $q+1$. Note also that π_0 is a morphism of degree $\deg \pi_0 = q+1$. We can therefore verify the Hurwitz formula [Har, Chapter IV, Corollary 2.4]

$$2\mathbf{g}(\overline{\mathbf{Y}}) - 2 = (\deg \pi_0)(2 \cdot \mathbf{g}(\mathbf{P}^1(\mathbb{F})) - 2) + \sum_{a \in \mathbf{P}^1(\mathbb{F}_q)} (e_a - 1),$$

as $\mathbf{g}(\mathbf{P}^1(\mathbb{F})) = 0$.

We will now extend the group $G \times \mu_{q+1}$ to a bigger group $\mathscr{G}$ still acting on $\mathbf{Y}$ (or $\overline{\mathbf{Y}}$). Set

$$\mathscr{G} = \{(g, \xi) \in \mathrm{GL}_2(\mathbb{F}_q) \times \mathbb{F}_{q^2}^\times \mid \det(g) = \xi^{1+q}\}.$$

It is then straightforward to verify that,

(2.5.2) *if* $(g, \xi) \in \mathscr{G}$ *and* $(x, y) \in \mathbf{Y}$, *then* $g \cdot (\xi x, \xi y) \in \mathbf{Y}$.

This defines for us an action of $\mathscr{G}$ on $\mathbf{Y}$ which extends naturally to an action on $\overline{\mathbf{Y}}$. Set

$$\mathscr{D} = \begin{cases} \langle(-\mathrm{I}_2, -1)\rangle & \text{if } q \equiv 3 \mod 4, \\ \langle(\sqrt{-1}\,\mathrm{I}_2, -\sqrt{-1})\rangle & \text{if } q \equiv 1 \mod 4. \end{cases}$$

Then $\mathscr{D}$ is a central subgroup of $\mathscr{G}$ contained in the kernel of the action on $\mathbf{Y}$ (and on $\overline{\mathbf{Y}}$). Even better, we have the following.

Lemma 2.5.3. *The group* $\mathscr{G}/\mathscr{D}$ *acts faithfully on* $\mathbf{Y}$ *(and* $\overline{\mathbf{Y}}$*).*

Proof. Let (g, ξ) be an element of $\mathscr{G}$ which acts trivially on $\mathbf{Y}$. Then (g, ξ) acts trivially on $\overline{\mathbf{Y}}$ (as $\mathbf{Y}$ is dense in $\overline{\mathbf{Y}}$) and, after passing to the quotient by $\{1\} \times \mu_{q+1}$ (which is a central subgroup of $\mathscr{G}$), we conclude that g acts trivially on on $\mathbf{P}^1(\mathbb{F})$ (by Theorem 2.4.2). Therefore g is a homothety: $g = \lambda \mathrm{I}_2$, with $\lambda \in \mathbb{F}_q^\times$.

Now, if $(x, y) \in \mathbf{Y}$, we have $(g, \xi) \cdot (x, y) = (x, y)$, that is $\lambda\xi = 1$. Therefore $\xi = \lambda^{-1}$. On the other hand, $\det(g) = \xi^{q+1}$, which implies that $\lambda^2 = \xi^{q+1}$ or, in other words, $\lambda^{q+3} = 1$. As $\lambda^{q-1} = 1$, we collude that $\lambda^4 = 1$, which finishes the proof. □

Let $\Delta = \mathscr{D} \cap (G \times \mu_{q+1}) = \langle(-\mathrm{I}_2, -1)\rangle$.

Corollary 2.5.4. *The group* $(G \times \mu_{q+1})/\Delta$ *acts faithfully on* $\mathbf{Y}$.

Denote by $p_1 : \mathscr{G} \to \mathsf{GL}_2(\mathbb{F}_q)$ and $p_2 : \mathscr{G} \to \mathbb{F}_{q^2}^\times$ the canonical projections, and $i_1 : \mu_{q+1} \to \mathscr{G}, \xi \mapsto (\mathsf{I}_2, \xi)$ and $i_2 : G \to \mathscr{G}, g \mapsto (g, 1)$. The group $G \times \mu_{q+1}$ is contained in $\mathscr{G}$ and we set $d : \mathscr{G} \to \mathbb{F}_q^\times, (g, \xi) \mapsto \det(g)$. We have a commutative diagram

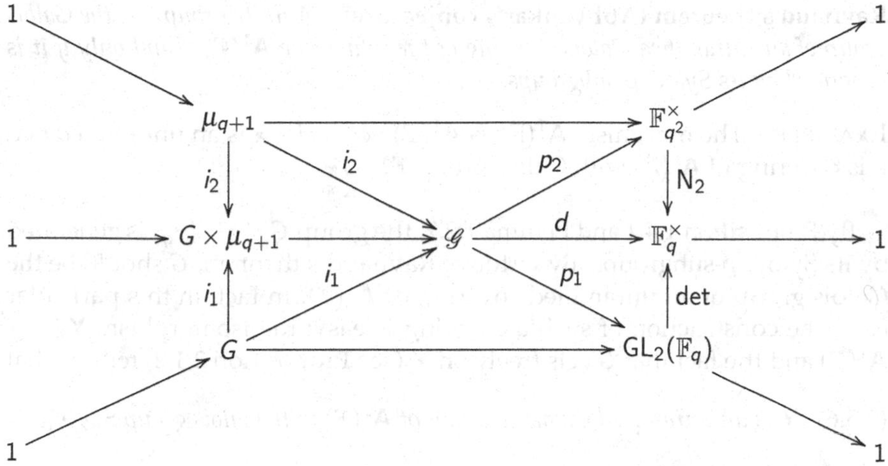

in which all straight lines of the form $1 \to \mathscr{X} \to \mathscr{G} \to \mathscr{Y} \to 1$ are exact sequences (which follows essentially from the surjectivity of N_2). In particular,

$$|\mathscr{G}| = q(q^2-1)^2. \tag{2.5.5}$$

It follows from Lemma 2.5.3 that

$$|\mathsf{Aut}\,\overline{\mathbf{Y}}| \geqslant \begin{cases} \dfrac{q(q^2-1)^2}{2} & \text{if } q \equiv 3 \mod 4, \\[2ex] \dfrac{q(q^2-1)^2}{4} & \text{if } q \equiv 1 \mod 4. \end{cases}$$

In particular, as soon as $q \geqslant 7$, we have, by 2.5.1,

$$|\mathsf{Aut}\,\overline{\mathbf{Y}}| > 84(\mathbf{g}(\overline{\mathbf{Y}}) - 1) = 42(q-2)(q+1).$$

This illustrates the fact that the "Hurwitz bound" [Har, Chapter IV, Exercise 2.5] is not valid in positive characteristic.

2.5.2. *Abhyankar's Conjecture (Raynaud's Theorem)**

It is not too difficult to show that if a finite group Γ is the Galois group of an unramified covering of the affine line $\mathbf{A}^1(\mathbb{F})$, then Γ is generated by its Sylow p-subgroups. The other implication was conjectured by Abhyankar and shown by Raynaud in a very difficult work [Ray].

Raynaud's theorem (Abhyankar's conjecture). *A finite group Γ is the Galois group of an unramified Galois covering of the affine line $\mathbf{A}^1(\mathbb{F})$ if and only if it is generated by its Sylow p-subgroups.*

EXAMPLE – The morphism $\mathbf{A}^1(\mathbb{F}) \to \mathbf{A}^1(\mathbb{F})$, $x \mapsto x^q - x$ is an unramified Galois covering of $\mathbf{A}^1(\mathbb{F})$ with Galois group $\mathbb{F}_q^+$. □

By Proposition 1.4.1 and Lemma 1.2.2, the group $G = \mathrm{SL}_2(\mathbb{F}_q)$ is generated by its Sylow p-subgroups. By virtue of Raynaud's theorem, G should be the Galois group of an unramified covering of $\mathbf{A}^1(\mathbb{F})$. In fact, in this particular case, the construction of such a covering is easy: the isomorphism $\mathbf{Y}/G \simeq \mathbf{A}^1(\mathbb{F})$ and the fact that G acts freely on $\mathbf{Y}$ (see Proposition 2.1.2) tells us that

(2.5.6) $\mathbf{Y}$ *is an unramified Galois covering of* $\mathbf{A}^1(\mathbb{F})$ *with Galois group* $\mathrm{SL}_2(\mathbb{F}_q)$.

Exercises

2.1. Show that the polynomial $XY^q - YX^q - 1$ in $\mathbb{F}[X, Y]$ is irreducible (***Hint:*** By performing the change of variables $(Z, T) = (X/Y, 1/Y)$ reduce the problem to showing that $T^{q+1} - Z^q - Z$ in $\mathbb{F}[Z, T]$ is irreducible. View this as a polynomial in T with coefficients $\mathbb{F}[Z]$ and use Eisenstein's criterion).

2.2*. Let $\mathbb{F}[X, Y]$ a the polynomial ring in two variables, which we identify with the algebra of polynomial functions on $\mathbf{A}^2(\mathbb{F})$. If $g \in G$, $P \in \mathbb{F}[X, Y]$ and $v \in \mathbf{A}^2(\mathbb{F})$, we set $(g \cdot P)(v) = P(g^{-1} \cdot v)$.

(a) Show that this does indeed give an action of G via $\mathbb{F}$-algebra automorphisms.
(b) Show that $XY^{q-1} - X^q$ and Y are algebraically independent and that $\mathbb{F}[X, Y]^U = \mathbb{F}[XY^{q-1} - X^q, Y]$.
(c) Show that $XY^q - YX^q$ divides $XY^{q^2} - YX^{q^2}$.
(d) Show that $D_1 = XY^q - YX^q$ and $D_2 = \dfrac{XY^{q^2} - YX^{q^2}}{XY^q - YX^q}$ are algebraically independent.
(e) Show that $\mathbb{F}[X, Y]^G = \mathbb{F}[D_1, D_2]$ (*Dickson invariants*).
(f) Use this to give another proof of Theorem 2.2.2.

2.3. Denote by Δ the subgroup of $G \times \mu_{q+1}$ generated by $(-\mathrm{I}_2, -1)$. The purpose of this exercise is to show that $(G \times \mu_{q+1})/\Delta$ does not act freely on $\mathbf{Y}$. To this end, choose $\xi \in \mu_{q+1} \setminus \{1, -1\}$ and let $v = (x, y) \in \mathbf{A}^2(\mathbb{F})$ be an eigenvector of $\mathbf{d}'(\xi)$ with eigenvalue ξ.

(a) Show that $xy^q - yx^q \neq 0$ (***Hint:*** $xy^q - yx^q = x\prod_{a \in \mathbb{F}_q}(y + ax)$).
(b) Let $\kappa \in \mathbb{F}^\times$ be such that $\kappa^{-1-q} = xy^q - yx^q$. Show that $\kappa v \in \mathbf{Y}$.
(c) Show that $(\mathbf{d}'(\xi), \xi^{-1})$ stabilises $\kappa v \in \mathbf{Y}$.

2.4. † Let $\mathbf{Z} = \{(x, y) \in \mathbf{A}^2(\mathbb{F}) \mid x^{q+1} + y^{q+1} + 1 = 0\}$. We keep the notation F for the restriction to $\mathbf{Z}$ of the Frobenius endomorphism F of $\mathbf{A}^2(\mathbb{F})$. The purpose of this exercise is to construct an isomorphism of $\mathbf{Y}$ and $\mathbf{Z}$ which commutes with F^4.

(a) Show that $\mathbf{Z}^{F^2} \neq \varnothing$. Deduce that there does not exist an isomorphism of varieties $\tau \colon \mathbf{Y} \xrightarrow{\sim} \mathbf{Z}$ such that $\tau \circ F^2 = F^2 \circ \tau$.

Let $z \in \mathbb{F}_{q^2} \setminus \mathbb{F}_q$ and $d \in \mathbb{F}$ be such that $d^{q+1} = -\dfrac{1}{z^q - z}$.

(b) Show that $d \in \mathbb{F}_{q^4}$.
(c) Let $g = \begin{pmatrix} d^q z^q & dz \\ d^q & d \end{pmatrix}$. Show that $g \in \mathrm{GL}_2(\mathbb{F}_{q^4})$ and that $g(\mathbf{Z}) = \mathbf{Y}$.

2.5. Denote by $\tau \colon \mathbf{Y}/U \to \mathbf{Y}/G$ the canonical projection. Set $\tau' = \bar{\gamma} \circ \tau \circ \bar{\upsilon}^{-1} \colon \mathbf{A}^1(\mathbb{F}) \setminus \{0\} \longrightarrow \mathbf{A}^1(\mathbb{F})$, so that the diagram

$$\begin{array}{ccc} \mathbf{Y}/U & \xrightarrow{\quad\tau\quad} & \mathbf{Y}/G \\ \Big\downarrow{\scriptstyle \bar{\upsilon}} & & \Big\downarrow{\scriptstyle \bar{\gamma}} \\ \mathbf{A}^1(\mathbb{F}) \setminus \{0\} & \xrightarrow{\quad\tau'\quad} & \mathbf{A}^1(\mathbb{F}) \end{array}$$

commutes. Show that $\tau'(y) = y^{-q}(y^{q^2} + y)$.

† The author is indebted to G. Lusztig to whom this exercise is due.

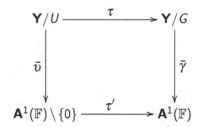

Part II
Ordinary Characters

The purpose of the next three chapters is to calculate the character table of $G = \mathrm{SL}_2(\mathbb{F}_q)$. To begin with, algebraic methods (in particular Harish-Chandra induction) give roughly half of the irreducible characters (see Chapter 3). The "other half" (the *cuspidal* characters) can be obtained in different ways (for example via ad hoc construction, as was done by Jordan [Jor] and Schur [Sch] in 1907). In this book we take the approach of constructing the cuspidal characters using the ℓ-adic cohomology of the Drinfeld curve (see Chapter 4). For this we will use results concerning ℓ-adic cohomology contained in Appendix A as well as some geometric properties obtained in Chapter 2.

Once the parametrisation has been obtained, the calculation of the character table is completed in Chapter 5: it raises geometric questions (trace on ℓ-adic cohomology) as well as more arithmetic questions (Gauss sums).

NOTATION – We fix a prime number ℓ different from p and K, an algebraic extension of the ℓ-adic field containing the $|\Gamma|$-th roots of unity, for all finite groups Γ encountered in this book. This ensures that the algebra $K\Gamma$ is split (by Brauer's theorem [Isa, Theorem 8.4]).

If G is a finite group, we will denote by $\mathrm{Irr}\,\Gamma$ its set of irreducible characters over K (which we identify with $\mathrm{Irr}\,K\Gamma$) and the Grothendieck group $\mathscr{K}_0(K\Gamma)$ will be identified with the character group $\mathbb{Z}\,\mathrm{Irr}\,\Gamma$. If M is a $K\Gamma$-module, we therefore identify its isomorphism class $[M]_\Gamma$ with its character. We will denote by reg_Γ the character of the regular representation $K\Gamma$ of Γ and by $\langle,\rangle_\Gamma$ the scalar product on $\mathscr{K}_0(K\Gamma)$ for which the set $\mathrm{Irr}\,\Gamma$ is an orthonormal basis. We denote by 1_Γ (or 1) the trivial character of Γ.

The group of linear characters of Γ (with values in $K^\times$) will be denoted by $\Gamma^\wedge$ (so that $\Gamma^\wedge \subseteq \mathrm{Irr}\,\Gamma$). Given $\alpha \in \Gamma^\wedge$, we denote by K_α the $K\Gamma$-module with underlying vector space K and on which Γ acts via the linear character α.

If $\chi \in \mathscr{K}_0(K\Gamma)$, we will denote by χ^* the virtual character defined by $\chi^*(\gamma) = \chi(\gamma^{-1})$ for all $\gamma \in \Gamma$: χ^* is the character *dual* to χ (and χ is *auto-dual* if $\chi = \chi^*$). If M is a left $K\Gamma$-module of finite type, its K-dual M^* can either be viewed as a right $K\Gamma$-module or as a left $K\Gamma$-module (via the inverse map); in the second case, we have $[M^*] = [M]^*$. If $\chi \in \mathrm{Irr}\,\Gamma$, we will denote by e_χ (or e_χ^Γ if necessary) the associated central primitive idempotent of $K\Gamma$:

$$e_\chi = \frac{\chi(1)}{|\Gamma|} \sum_{\gamma\in\Gamma} \chi(\gamma^{-1})\,\gamma.$$

Chapter 3
Harish-Chandra Induction

In this chapter we study Harish-Chandra induction, which associates to a T-module the G-module obtained by first extending the T-module to a B-module (letting U act trivially) and then inducing to G. This construction allows us to obtain roughly half of the irreducible characters of G.

3.1. Bimodules

Let Γ and Γ' be two finite groups and let M be a $(K\Gamma, K\Gamma')$-bimodule of finite type. The dual bimodule $M^* = \mathrm{Hom}(M, K)$ is naturally a $(K\Gamma', K\Gamma)$-bimodule. We define two functors

$$\begin{array}{rccc} \mathscr{F}_M\colon & K\Gamma'\text{-mod} & \longrightarrow & K\Gamma\text{-mod} \\ & V' & \longmapsto & M \otimes_{K\Gamma'} V' \end{array}$$

and

$$\begin{array}{rccc} {}^*\mathscr{F}_M\colon & K\Gamma\text{-mod} & \longrightarrow & K\Gamma'\text{-mod} \\ & V & \longmapsto & M^* \otimes_{K\Gamma} V. \end{array}$$

Because $K\Gamma$ and $K\Gamma'$ are semi-simple algebras, the bimodule M is projective both as a left $K\Gamma$-module and as a right $K\Gamma'$-module. It follows that the functors $\mathscr{F}_M$ and ${}^*\mathscr{F}_M$ are left and right adjoint:

(3.1.1) $$\mathrm{Hom}_{K\Gamma}(V, \mathscr{F}_M V') \simeq \mathrm{Hom}_{K\Gamma'}({}^*\mathscr{F}_M V, V')$$

and

(3.1.2) $$\mathrm{Hom}_{K\Gamma}(\mathscr{F}_M V', V) \simeq \mathrm{Hom}_{K\Gamma'}(V', {}^*\mathscr{F}_M V)$$

We denote by $\mathsf{F}_M\colon \mathscr{K}_0(K\Gamma') \to \mathscr{K}_0(K\Gamma)$ and ${}^*\mathsf{F}_M\colon \mathscr{K}_0(K\Gamma) \to \mathscr{K}_0(K\Gamma')$ the $\mathbb{Z}$-linear maps induced by $\mathscr{F}_M$ and ${}^*\mathscr{F}_M$ respectively. They are characterised

C. Bonnafé, *Representations of* $\mathrm{SL}_2(\mathbb{F}_q)$, Algebra and Applications 13,
DOI 10.1007/978-0-85729-157-8_3, © Springer-Verlag London Limited 2011

by

$$\mathrm{F}_M[V']_{\Gamma'} = [M \otimes_{K\Gamma'} V']_\Gamma \quad \text{and} \quad {}^*\mathrm{F}_M[V]_\Gamma = [M^* \otimes_{K\Gamma} V]_{\Gamma'}. \tag{3.1.3}$$

Recall that, if V_1 and V_2 are two $K\Gamma$-modules of finite type, then

$$\langle [V_1], [V_2] \rangle_\Gamma = \dim_K \mathrm{Hom}_{K\Gamma}(V_1, V_2). \tag{3.1.4}$$

Recall also that the Grothendieck group $\mathscr{K}_0(K\Gamma)$ is identified with the group of (virtual) characters of Γ. Under this identification, it follows from 3.1.1 and 3.1.4 that

$$\langle \chi, \mathrm{F}_M(\chi') \rangle_\Gamma = \langle {}^*\mathrm{F}_M(\chi), \chi' \rangle_{\Gamma'} \tag{3.1.5}$$

for all $\chi \in \mathscr{K}_0(K\Gamma)$ and $\chi' \in \mathscr{K}_0(K\Gamma')$. If $(\gamma, \gamma') \in \Gamma \times \Gamma'$, we denote by $\mathrm{Tr}_M(\gamma, \gamma')$ the trace of (γ, γ') on M. We have (see, for example, [DiMi, Proposition 4.5])

$$\mathrm{F}_M \chi'(\gamma) = \frac{1}{|\Gamma'|} \sum_{\gamma' \in \Gamma'} \mathrm{Tr}_M(\gamma, \gamma')\, \chi'(\gamma'^{-1}) \tag{3.1.6}$$

and

$${}^*\mathrm{F}_M \chi(\gamma') = \frac{1}{|\Gamma|} \sum_{\gamma \in \Gamma} \mathrm{Tr}_M(\gamma, \gamma')\, \chi(\gamma^{-1}). \tag{3.1.7}$$

3.2. Harish-Chandra Induction

3.2.1. Definition

In our group G, Harish-Chandra induction is defined using the bimodule $K[G/U]$: by convention, $K[G/U]$ is the K-vector space with basis G/U on which G (respectively T) acts by left (respectively right) translations on the basis vectors. This is well-defined as T normalizes U. The dual of $K[G/U]$ may be naturally identified with $K[U\backslash G]$, with the action of the group G (respectively T) given by right (respectively left) translations. In this way we obtain two functors

$$\begin{array}{rccc} \mathscr{R}_K : & KT\text{-mod} & \longrightarrow & KG\text{-mod} \\ & V & \longmapsto & K[G/U] \otimes_{KT} V \end{array}$$

and

$$\begin{array}{rccc} {}^*\mathscr{R}_K : & KG\text{-mod} & \longrightarrow & KT\text{-mod} \\ & W & \longmapsto & K[U\backslash G] \otimes_{KG} W, \end{array}$$

called *Harish-Chandra induction* and *restriction* respectively. We denote by $\mathrm{R}\colon \mathscr{K}_0(KT) \to \mathscr{K}_0(KG)$ and $^*\mathrm{R}\colon \mathscr{K}_0(KG) \to \mathscr{K}_0(KT)$, the induced $\mathbb{Z}$-linear maps.

An irreducible character g of G will be called *cuspidal* if there does not exist a character α of $T \simeq \mu_{q-1}$ for which $\langle \gamma, \mathrm{R}(\alpha)\rangle_G \neq 0$. In other words, taking into account 3.1.5,

(3.2.1) γ *is cuspidal if and only if* $^*\mathrm{R}(\gamma) = 0.$

3.2.2. Other Constructions

In order to study these functors, it will be useful to interpret them using classical induction and restriction. We will need the following notation: if V (respectively α) is a KT-module (respectively a character of KT), we will denote by V_B (respectively α_B) its "restriction" to B via the natural projection $B \to T$. Note that $[V_B]_B = ([V]_T)_B$.

Proposition 3.2.2. *Let V (respectively W) be a KT-module (respectively a KG-module). Then*

$$\mathscr{R}_K V \simeq \mathrm{Ind}_B^G V_B \quad \text{and} \quad {}^*\mathscr{R}_K W \simeq W^U.$$

Proof. We have $\mathrm{Ind}_B^G V_B = KG \otimes_{KB} V_B$ and $\mathscr{R}_K(V) = K[G/U] \otimes_{KT} V$. Set

$$e_U = \frac{1}{|U|} \sum_{u \in U} u.$$

Denote by $\tau_U\colon KG \to K[G/U]$ the canonical morphism and $\sigma_U\colon K[G/U] \to KG$, $gU \mapsto ge_U$. We have

$$\sigma_U \circ \tau_U(g) = ge_U \quad \text{and} \quad \tau_U \circ \sigma_U(x) = x.$$

Set

$$\begin{aligned} \varphi\colon\ KG \otimes_{KB} V_B &\longrightarrow K[G/U] \otimes_{KT} V \\ a \otimes_{KB} v &\longmapsto \tau_U(a) \otimes_{KT} v \end{aligned}$$

and

$$\begin{aligned} \psi\colon\ K[G/U] \otimes_{KT} V &\longrightarrow KG \otimes_{KB} V_B \\ a \otimes_{KT} v &\longmapsto \sigma_U(a) \otimes_{KB} v. \end{aligned}$$

It is easy to verify that φ and ψ are well-defined and morphisms of KG-modules. It is also clear that $\varphi \circ \psi = \mathrm{Id}_{K[G/U] \otimes_{KT} V}$. In order to show that $\psi \circ \varphi = \mathrm{Id}_{KG \otimes_{KB} V_B}$, it is enough to note that $a \otimes_{KB} v = ae_U \otimes_{KB} v$ for all $a \in KG$ and $v \in V_B$. This shows the first isomorphism.

Note that $^*\mathscr{R}_K W = K[U \backslash G] \otimes_{KG} W$. Set

$$\begin{aligned} \varphi' : \; K[U\backslash G] \otimes_{KG} W &\longrightarrow W^U \\ a \otimes_{KG} w &\longmapsto \sigma_U^*(a) w \end{aligned}$$

and

$$\begin{aligned} \psi' : \; W^U &\longrightarrow K[U\backslash G] \otimes_{KG} W \\ w &\longmapsto U \otimes_{KG} w. \end{aligned}$$

Here, $\sigma_U^*(Ug) = e_U g$ for all $g \in G$. It is easy to verify that φ' and ψ' are well-defined and are mutually inverse isomorphisms of KT-modules. □

Corollary 3.2.3. *Let α be a character of T. Then* $\mathsf{R}(\alpha) = \mathsf{Ind}_B^G \alpha_B$.

3.2.3. *Mackey Formula*

It follows easily from the Bruhat decomposition $G = B \dot\cup BsB$ (see 1.1.4 and 1.1.5), from Corollary 3.2.3 and from the Mackey formula for classical induction and restriction that, if $\alpha, \beta \in \mathscr{K}_0(KT)$, then

$$\langle \mathsf{R}(\alpha), \mathsf{R}(\beta) \rangle_G = \langle \alpha, \beta \rangle_T + \langle \alpha, {}^s\beta \rangle_T.$$

Here, ${}^s\beta(t) = \beta(s^{-1}ts)$. As $s^{-1}ts = t^{-1}$ for all $t \in T$, we have, if $\alpha, \beta \in \mathscr{K}_0(KT)$,

(3.2.4) $$\langle \mathsf{R}(\alpha), \mathsf{R}(\beta) \rangle_G = \langle \alpha, \beta \rangle_T + \langle \alpha, \beta^* \rangle_T.$$

Note that, if β is a linear character, then $\beta^* = \beta^{-1}$.

Let us now fix $\alpha \in T^\wedge = \mathsf{Irr}\, T$. By 3.2.4, we have:

- If $\alpha^2 \neq 1$, then $\mathsf{R}(\alpha) = \mathsf{R}(\alpha^{-1}) \in \mathsf{Irr}\, G$.
- If $\alpha = \alpha_0$ (where α_0 is the unique linear character of order 2 of $T \simeq \mu_{q-1}$, which exists because q is odd), then

  $$\mathsf{R}(\alpha_0) = \mathsf{R}_+(\alpha_0) + \mathsf{R}_-(\alpha_0),$$

  where $\mathsf{R}_\pm(\alpha_0) \in \mathsf{Irr}\, G$ and $\mathsf{R}_+(\alpha_0) \neq \mathsf{R}_-(\alpha_0)$.
- 1_G is a factor of $\mathsf{R}(1)$ and we denote by St_G the other irreducible factor of $\mathsf{R}(1)$ (*the Steinberg character*). We have

  $$\mathsf{R}(1) = 1_G + \mathsf{St}_G, \quad \text{with } \deg 1_G = 1 \text{ and } \deg \mathsf{St}_G = q.$$

- If $\alpha \notin \{\beta, \beta^{-1}\}$, then $\langle \mathsf{R}(\alpha), \mathsf{R}(\beta) \rangle_G = 0$.

3.2.4. *Restriction from* $\mathrm{GL}_2(\mathbb{F}_q)$

In order to calculate the degrees of the characters $R_+(\alpha_0)$ and $R_-(\alpha_0)$ we will show that they are conjugate under the action of $\widetilde{G} = \mathrm{GL}_2(\mathbb{F}_q)$. Denote by $\widetilde{B}$ (respectively $\widetilde{T}$) the subgroup of $\widetilde{G}$ consisting of upper triangular (respectively diagonal) matrices. Then

(3.2.5) $$\widetilde{B} = \widetilde{T} \ltimes U \quad \text{and} \quad \widetilde{G} = \widetilde{B} \,\dot\cup\, \widetilde{B}s\widetilde{B}.$$

If $\tilde{\alpha}$ is a character of $\widetilde{T}$, we denote by $\tilde{\alpha}_{\widetilde{B}}$ its "restriction" to $\widetilde{B}$ via the surjection $\widetilde{B} \to \widetilde{T}$. We have

$$\mathrm{Res}_B^{\widetilde{B}}\, \tilde{\alpha}_{\widetilde{B}} = (\mathrm{Res}_T^{\widetilde{T}})_B.$$

We set

$$\widetilde{R}(\tilde{\alpha}) = \mathrm{Ind}_{\widetilde{B}}^{\widetilde{G}}\, \tilde{\alpha}_{\widetilde{B}}.$$

As $\widetilde{G} = G \cdot \widetilde{B}$, we deduce from the Mackey formula for classical induction and restriction that

(3.2.6) $$\mathrm{Res}_G^{\widetilde{G}}\, \widetilde{R}(\tilde{\alpha}) = R(\mathrm{Res}_T^{\widetilde{T}}\, \tilde{\alpha}).$$

On the other hand, it follows from 3.2.5 that

(3.2.7) $$\langle \widetilde{R}(\tilde{\alpha}), \widetilde{R}(\tilde{\beta}) \rangle = \langle \tilde{\alpha}, \tilde{\beta} \rangle_{\widetilde{T}} + \langle \tilde{\alpha}, {}^s\tilde{\beta} \rangle_{\widetilde{T}}.$$

In particular,

(3.2.8) *if $\tilde{\alpha} \in \widetilde{T}^\wedge$ is such that ${}^s\tilde{\alpha} \neq \tilde{\alpha}$, then $\widetilde{R}(\tilde{\alpha})$ is irreducible.*

The following lemma is elementary.

Lemma 3.2.9. *If $\tilde{\alpha}_0$ it an extension to $\widetilde{T}$ of α_0, then ${}^s\tilde{\alpha}_0 \neq \tilde{\alpha}_0$.*

The next corollary then follows immediately from 3.2.6 and 3.2.8.

Corollary 3.2.10. *If $\tilde{\alpha}_0$ is an extension to $\widetilde{T}$ of α_0, then $\widetilde{R}(\tilde{\alpha}_0)$ is an irreducible character of $\widetilde{G}$.*

By 3.2.6 and Clifford theory [Isa, Theorem 6.2], we obtain the following.

Corollary 3.2.11. *The irreducible characters $R_+(\alpha_0)$ and $R_-(\alpha_0)$ of G are conjugate under the action of $\widetilde{G}$.*

In particular,

(3.2.12) $$\deg R_+(\alpha_0) = \deg R_-(\alpha_0) = \frac{q+1}{2}.$$

Another proof of 3.2.12 will be given in Exercise 3.3.

3.2.5. Summary

We have therefore obtained:

- one linear character 1_G;
- one character of degree q, the Steinberg character St_G;
- $(q-3)/2$ characters of degree $q+1$ (the characters $\mathrm{R}(\alpha)$, $\alpha^2 \neq 1$);
- two characters $\mathrm{R}_+(\alpha_0)$ and $\mathrm{R}_-(\alpha_0)$ of degree $(q+1)/2$ (and conjugate under $\widetilde{G}$).

This yields $(q+5)/2$ irreducible characters. As the number of conjugacy classes (and therefore the number of irreducible characters) of G is $q+4$, it remains to construct $(q+3)/2$ irreducible characters: these are the cuspidal characters of G. Note also that

$$\mathrm{R}(\mathrm{reg}_T) = \mathrm{Ind}_U^G 1_U = \underbrace{1_G + \mathrm{St}_G}_{\mathrm{R}(1)} + \underbrace{\mathrm{R}_+(\alpha_0) + \mathrm{R}_-(\alpha_0)}_{\mathrm{R}(\alpha_0)} + \sum_{\alpha^2 \neq 1} \mathrm{R}(\alpha). \tag{3.2.13}$$

Exercises

3.1. Let $\widetilde{Z}$ denote the centre of $\widetilde{G}$.

(a) Show that $\widetilde{Z} = \{\mathrm{diag}(a,a) \mid a \in \mathbb{F}_q^\times\}$.
(b) Show that $|\widetilde{G}/G \cdot \widetilde{Z}| = 2$.

3.2. Use the fact that $\mathrm{R}(1)$ is the character of the permutation representation $K[\mathbf{P}^1(\mathbb{F}_q)]$ in order to calculate its value on all conjugacy classes. Then verify that, if $g \in G$,

$$\mathrm{St}_G(g) = \begin{cases} |\mathrm{C}_G(g)|_p & \text{if } g \text{ is semi-simple,} \\ 0 & \text{otherwise.} \end{cases}$$

Here, $|\Gamma|_p$ denotes the largest power of p which divides $|\Gamma|$.

Denote by $\mathrm{St}_{\widetilde{G}} = \widetilde{\mathrm{R}}(1) - 1_{\widetilde{G}}$. Show that $\mathrm{St}_{\widetilde{G}}$ is an irreducible character of $\widetilde{G}$, that $\mathrm{Res}_G^{\widetilde{G}} \mathrm{St}_{\widetilde{G}} = \mathrm{St}_G$ and that, if $g \in \widetilde{G}$, then

$$\mathrm{St}_{\widetilde{G}}(g) = \begin{cases} |\mathrm{C}_{\widetilde{G}}(g)|_p & \text{if } g \text{ is semi-simple,} \\ 0 & \text{otherwise.} \end{cases}$$

3.3* The goal of this exercise is to give a new proof of 3.2.12. If $g \in G$, we set

$$\mathscr{F}(gU) = \sum_{u \in U} gusU \in K[G/U].$$

(a) Show that this gives a well-defined K-linear endomorphism $\mathscr{F}$ of $K[G/U]$.

(b) Show that, if $g \in G$, $x \in K[G/U]$ and $t \in T$, then

$$\mathscr{F}(g \cdot x \cdot t) = g \cdot \mathscr{F}(x) \cdot {}^s t.$$

If $\alpha \in T^\wedge$, we let $V_\alpha = \mathscr{R}_K(K_\alpha)$ $(= K[G/U] \otimes_{KT} K_\alpha)$.

(c) Show that $\mathscr{F}$ induces an isomorphism of KG-modules $V_\alpha \simeq V_{{}^s\alpha}$.

Now fix $\alpha \in T^\wedge$ such that $\alpha^2 = 1$. By (c), $\mathscr{F}$ induces an automorphism of the KG-module V_α which we will denote by $\mathscr{F}_\alpha$.

(d) Show that $\mathscr{F}_\alpha^2 = \alpha(-1)q \operatorname{Id}_{V_\alpha} + \Big(\sum_{a \in \mathbb{F}_q^\times} \alpha(a)\Big)\mathscr{F}_\alpha$ and that $\operatorname{Tr}(\mathscr{F}_\alpha) = 0$.

(e) Use (d) to deduce that $(\mathscr{F}_1 - q \operatorname{Id}_{V_1})(\mathscr{F}_1 + \operatorname{Id}_{V_1}) = 0$ and that, if we set $I = \operatorname{Ker}(\mathscr{F}_1 - q \operatorname{Id}_{V_1})$ and $S = \operatorname{Ker}(\mathscr{F}_1 + \operatorname{Id}_{V_1})$, then $V_1 = I \oplus S$ is a decomposition of V_1 as a sum of irreducible KG-modules such that $[I] = 1_G$ and $[S] = \operatorname{St}_G$.

(f) Deduce from (d) that, if we set $V_{\alpha_0}^+ = \operatorname{Ker}(\mathscr{F}_{\alpha_0} - \sqrt{\alpha_0(-1)q} \operatorname{Id}_{V_{\alpha_0}})$ and $V_{\alpha_0}^- = \operatorname{Ker}(\mathscr{F}_{\alpha_0} + \sqrt{\alpha_0(-1)q} \operatorname{Id}_{V_{\alpha_0}})$, then $V_{\alpha_0} = V_{\alpha_0}^+ \oplus V_{\alpha_0}^-$ is a decomposition of V_{α_0} as a sum of irreducible KG-modules and that $\dim_K V_{\alpha_0}^\pm = \dfrac{q+1}{2}$.

REMARK – We have

$$\alpha_0(-1) = \begin{cases} 1 & \text{if } q \equiv 1 \mod 4, \\ -1 & \text{if } q \equiv 3 \mod 4. \end{cases}$$

Consequently, the number $\alpha_0(-1)$ is characterised by the property that $\alpha_0(-1)q \equiv 1 \mod 4$. Note that this implies that $\dfrac{1 \pm \sqrt{\alpha_0(-1)q}}{2}$ is an algebraic integer.

3.4. Let V be a finite dimensional left KT-module. We view the dual V^* as a left KT-module. Show that we have an isomorphism of KG-modules $\mathscr{R}_K V^* \simeq (\mathscr{R}_K V)^*$. Hence deduce that $\operatorname{R}(\alpha^*) = \operatorname{R}(\alpha)^*$ for every character α of T.

In particular, $\operatorname{R}(\alpha_0)^* = \operatorname{R}(\alpha_0)$. Show that

$$\operatorname{R}_+(\alpha_0)^* = \begin{cases} \operatorname{R}_+(\alpha_0) & \text{if } q \equiv 1 \mod 4, \\ \operatorname{R}_-(\alpha_0) & \text{if } q \equiv 3 \mod 4. \end{cases}$$

Hint: Use Exercise 3.3 (and the remark that follows it, which implies that $\sqrt{\alpha_0(-1)q}$ is a *real* number if and only if $q \equiv 1 \mod 4$).

Chapter 4
Deligne-Lusztig Induction

We will use the action of $G \times \mu_{q+1}$ on $\mathbf{Y}$ to construct a morphism between the Grothendieck groups $\mathscr{K}_0(K\mu_{q+1})$ and $\mathscr{K}_0(KG)$. To this end, from now on we will view the monoid $\mu_{q+1} \rtimes \langle F \rangle_{\mathrm{mon}}$ as acting *on the right* on the Drinfeld curve $\mathbf{Y}$. It follows that the cohomology groups $H_c^i(\mathbf{Y})$ inherit the structure of $(KG, K[\mu_{q+1} \rtimes \langle F \rangle_{\mathrm{mon}}])$-bimodules.

We will systematically use the results of Appendix A (which are referenced as A.x.y).

4.1. Definition and First Properties

4.1.1. *Definition*

If θ is a character of μ_{q+1}, we set

$$\mathrm{R}'(\theta) = -\sum_{i \geqslant 0} (-1)^i \, [H_c^i(\mathbf{Y}) \otimes_{K\mu_{q+1}} V_\theta]_G,$$

where V_θ is a $K\mu_{q+1}$-module admitting the character θ. If θ is linear, we take $V_\theta = K_\theta$. This defines a $\mathbb{Z}$-linear map

$$\mathrm{R}' \colon \mathscr{K}_0(K\mu_{q+1}) \longrightarrow \mathscr{K}_0(KG)$$

which sends a character of μ_{q+1} to a *virtual* character of G. The linear map R' will be called *Deligne-Lusztig induction.*

As the curve $\mathbf{Y}$ is affine and irreducible of dimension 1 it follows from Theorem A.2.1(b) that $H_c^i(\mathbf{Y}) = 0$ if $i \notin \{1, 2\}$. As a consequence,

$$\mathrm{R}'(\theta) = [H_c^1(\mathbf{Y}) \otimes_{K\mu_{q+1}} V_\theta]_G - [H_c^2(\mathbf{Y}) \otimes_{K\mu_{q+1}} V_\theta]_G.$$

DOI 10.1007/978-0-85729-157-8_4,

On the other hand, the irreducibility of $\mathbf{Y}$ (see Proposition 2.1.1) and Theorem A.2.1(c) tells us that

$$[H_c^2(\mathbf{Y})]_{G\times\mu_{q+1}} = 1_{G\times\mu_{q+1}}. \tag{4.1.1}$$

The following result is then immediate.

Corollary 4.1.2. *If θ is a non-trivial linear character of μ_{q+1}, then*

$$\mathsf{R}'(\theta) = [H_c^1(\mathbf{Y}) \otimes_{K\mu_{q+1}} V_\theta]_G.$$

In this case, $\mathsf{R}'(\theta)$ is a character of G.

Proposition 4.1.3. *Let $\theta \in \mathscr{K}_0(K\mu_{q+1})$. Then $\mathsf{R}'(\theta) = \mathsf{R}'(\theta^*) = \mathsf{R}'(\theta)^*$.*

Proof. Denote by $\varphi\colon \mu_{q+1} \to \mu_{q+1}$ the homomorphism $\xi \mapsto \xi^{-1}$. In order to show the first equality, it suffices to show that $\mathsf{R}'(\theta) = \mathsf{R}'({}^\varphi\theta)$, where ${}^\varphi\theta(\xi) = \theta(\varphi(\xi))$. Denote by ${}^\varphi H_c^i(\mathbf{Y})$ the $(KG, K\mu_{q+1})$-bimodule on which the action of μ_{q+1} is twisted by φ. It is enough to show that the bimodules $H_c^i(\mathbf{Y})$ and ${}^\varphi H_c^i(\mathbf{Y})$ are isomorphic. However, the endomorphism F of $\mathbf{Y}$ induces an K-linear automorphism of $H_c^i(\mathbf{Y})$ (see Theorem A.2.7(c)). It is easy to see (taking account of the commutation relations between F and the elements of G and μ_{q+1}) that F induces an isomorphism of $(KG, K\mu_{q+1})$-bimodules $H_c^i(\mathbf{Y}) \simeq {}^\varphi H_c^i(\mathbf{Y})$. The first equality is therefore proven.

We now turn to the second equality. Let $g \in G$. Then, by 3.1.6, we have

$$\mathsf{R}'(\theta)(g) = \frac{1}{q+1} \sum_{\xi\in\mu_{q+1}} \mathsf{Tr}^*_{\mathbf{Y}}(g, \xi^{-1})\, \theta(\xi).$$

Therefore

$$\mathsf{R}'(\theta)^*(g) = \frac{1}{q+1} \sum_{\xi\in\mu_{q+1}} \mathsf{Tr}^*_{\mathbf{Y}}(g^{-1}, \xi^{-1})\, \theta^*(\xi^{-1}).$$

Now, by A.2.5, we have $\mathsf{Tr}^*_{\mathbf{Y}}(g, \xi) \in \mathbb{Z}$, therefore $\mathsf{Tr}^*_{\mathbf{Y}}(g, \xi) = \mathsf{Tr}^*_{\mathbf{Y}}(g^{-1}, \xi^{-1})$. It follows that

$$\mathsf{R}'(\theta)^*(g) = \frac{1}{q+1} \sum_{\xi\in\mu_{q+1}} \mathsf{Tr}^*_{\mathbf{Y}}(g, \xi)\, \theta^*(\xi^{-1}) = \mathsf{R}'(\theta^*)(g),$$

as expected. □

4.1.2. *The Character* $\mathsf{R}'(1)$

We have, by definition,

$$\mathsf{R}'(1) = [H_c^1(\mathbf{Y})^{\mu_{q+1}}]_G - [H_c^2(\mathbf{Y})^{\mu_{q+1}}]_G.$$

Therefore, by 4.1.1 and A.2.3,

$$\mathrm{R}'(1) - [H^1_c(\mathbf{Y}/\mu_{q+1})]_G - 1_G.$$

Now $\mathbf{Y}/\mu_{q+1} \simeq \mathbf{P}^1(\mathbb{F}) \setminus \mathbf{P}^1(\mathbb{F}_q)$ (see Theorem 2.2.4). Therefore, by Theorem A.2.6(a),

$$H^*_c(\mathbf{Y}/\mu_{q+1})_G = H^*_c(\mathbf{P}^1(\mathbb{F}))_G - H^*_c(\mathbf{P}^1(\mathbb{F}_q))_G.$$

Now, $H^*_c(\mathbf{P}^1(\mathbb{F}_q))_G = [K[G/B]]_G = 1_G + \mathrm{St}_G$ and $H^*_c(\mathbf{P}^1(\mathbb{F})) = 2 \cdot 1_G$ (see A.3.1 and A.3.2). Hence

(4.1.4) $$\mathrm{R}'(1) = \mathrm{St}_G - 1_G$$

and therefore

(4.1.5) $$[H^i_c(\mathbf{Y}/\mu_{q+1})]_G = \begin{cases} 1_G & \text{if } i = 2, \\ \mathrm{St}_G & \text{if } i = 1, \\ 0 & \text{otherwise.} \end{cases}$$

Although we have succeeded in calculating the character $\mathrm{R}'(1)$, we still have not obtained any new characters of G.

4.1.3. Dimensions

Let ξ be a non-trivial element of μ_{q+1}. Then, by Theorem A.2.6(d),

$$\mathrm{Tr}^*_{\mathbf{Y}}(\xi) = \mathrm{Tr}^*_{\mathbf{Y}^\xi}(1).$$

However $\mathbf{Y}^\xi = \varnothing$, therefore

$$\mathrm{Tr}^*_{\mathbf{Y}}(\xi) = 0.$$

We deduce from this that, as a character of μ_{q+1}, $H^*_c(\mathbf{Y})$ is a multiple of the character of the regular representation. So, for all $\theta \in (\mu_{q+1})^\wedge$,

(4.1.6) $$\deg \mathrm{R}'(\theta) = \deg \mathrm{R}'(1) = q - 1.$$

4.1.4. Cuspidality

The goal of this subsection is to show that, as soon as θ is a non-trivial linear character of μ_{q+1}, the irreducible components of $\mathrm{R}'(\theta)$ are cuspidal. We first start by a result which shows that Harish-Chandra induction is orthogonal to Deligne-Lusztig induction.

Theorem 4.1.7. *If α and θ are characters of $T \simeq \mu_{q-1}$ and μ_{q+1} respectively, then*

$$\langle \mathsf{R}(\alpha), \mathsf{R}'(\theta)\rangle_G = 0.$$

Proof. We may suppose that θ is a linear character. We have $\mathsf{R}'(1) = -1_G + \mathsf{St}_G$, $\mathsf{R}(1) = 1_G + \mathsf{St}_G$ and therefore $\langle \mathsf{R}(1), \mathsf{R}'(1)\rangle_G = 0$. On the other hand, if α is a character of T not containing 1 as an irreducible factor then $\langle \mathsf{R}(\alpha), \mathsf{R}'(1)\langle_G = 0$.

We may therefore suppose that $\theta \neq 1$. In this case, $\mathsf{R}'(\theta) = [H^1_c(\mathbf{Y}) \otimes_{KT} K_\theta]_G$ is a character (not only a virtual character) of G, therefore it suffices to show the result when $\alpha = \mathrm{reg}_T$. But,

$$\langle \mathsf{R}'(\theta), \mathsf{R}(\mathrm{reg}_T)\rangle_G = \langle \mathsf{R}'(\theta), \mathrm{Ind}_U^G 1_U\rangle_G = \dim_K(H^1_c(\mathbf{Y})^U \otimes_{K\mu_{q+1}} K_\theta).$$

By A.2.3, we have

$$H^1_c(\mathbf{Y})^U = H^1_c(\mathbf{Y}/U) = H^1_c(\mathbf{A}^1(\mathbb{F}) \setminus \{0\}),$$

with the last equality following from Theorem 2.2.3. But, by A.3.4 and A.3.5, we have

$$[H^1_c(\mathbf{A}^1(\mathbb{F}) \setminus \{0\})]_{\mu_{q+1}} = 1_{\mu_{q+1}},$$

and therefore $\dim_K(H^1_c(\mathbf{Y})^U \otimes_{K\mu_{q+1}} K_\theta) = 0$ as θ is a non-trivial linear character. $\square$

Theorem 4.1.7 shows us that, if θ is a non-trivial linear character of μ_{q+1}, then the irreducible components of $\mathsf{R}'(\theta)$ are *cuspidal*. Moreover, equality 4.1.6 shows that $\mathsf{R}'(\theta) \neq 0$. In order to obtain new irreducible characters of G all that remains is to decompose $\mathsf{R}'(\theta)$.

4.2. Mackey Formula

The goal of this section is to prove the following theorem (compare 3.2.4).

Mackey formula. *Let θ and η be two elements of $\mathcal{K}_0(K\mu_{q+1})$. We have*

$$\langle \mathsf{R}'(\theta), \mathsf{R}'(\eta)\rangle_G = \langle \theta, \eta\rangle_{\mu_{q+1}} + \langle \theta, \eta^*\rangle_{\mu_{q+1}}. \tag{4.2.1}$$

REMARK – If η is a linear character of μ_{q+1}, then $\eta^* = \eta^{-1}$. $\square$

The rest of this section is dedicated to the proof of this Mackey formula. We begin with a crucial geometric result determining the class of the module $H^*_c((\mathbf{Y} \times \mathbf{Y})/G)$ in the Grothendieck group of $\mu_{q+1} \times \mu_{q+1}$-modules. We will need the following notation. We denote by $\mu^{(1)}_{q+1}$ (respectively $\mu^{(2)}_{q+1}$) the $\mu_{q+1} \times \mu_{q+1}$-set with underlying set μ_{q+1} and such that, if $\zeta \in \mu_{q+1}$ and

$(\xi,\xi')\in\mu_{q+1}\times\mu_{q+1}$, then $(\xi,\xi')\cdot\zeta=\xi\xi'\zeta$ (respectively $(\xi,\xi')\cdot\zeta=\xi^{-1}\xi'\zeta$). This defines two permutation $K(\mu_{q+1}\times\mu_{q+1})$-modules. Then

(4.2.2) $$H^*_c((\mathbf{Y}\times\mathbf{Y})/G)_{\mu_{q+1}\times\mu_{q+1}}=[K[\mu^{(1)}_{q+1}]]_{\mu_{q+1}\times\mu_{q+1}}+[K[\mu^{(2)}_{q+1}]]_{\mu_{q+1}\times\mu_{q+1}}.$$

Proof (of 4.2.2). Set

$$\mathbf{Z}=\mathbf{Y}\times\mathbf{Y}=\{(x,y,z,t)\in\mathbf{A}^4(\mathbb{F})\mid xy^q-yx^q=1\text{ and }zt^q-tz^q=1\}.$$

Define

$$\mathbf{Z}_0=\{(x,y,z,t)\in\mathbf{Z}\mid xt-yz=0\}$$

and
$$\mathbf{Z}_{\neq 0}=\{(x,y,z,t)\in\mathbf{Z}\mid xt-yz\neq 0\}.$$

Then $\mathbf{Z}_0$ and $\mathbf{Z}_{\neq 0}$ are $(G\times\mu_{q+1}\times\mu_{q+1})$-stable subvarieties of $\mathbf{Z}$. By Theorem A.2.6(a), we therefore have

(α) $$[H^*_c(\mathbf{Z}/G)]_{\mu_{q+1}\times\mu_{q+1}}=[H^*_c(\mathbf{Z}_0/G)]_{\mu_{q+1}\times\mu_{q+1}}+[H^*_c(\mathbf{Z}_{\neq 0}/G)]_{\mu_{q+1}\times\mu_{q+1}}.$$

It is enough to show that the two terms on the right of (α) correspond to the two terms on the right of 4.2.2.

We begin with the first term. It is very easy to see that the morphism

$$\begin{array}{ccc}\mu_{q+1}\times\mathbf{Y} & \longrightarrow & \mathbf{Z}_0\\ (\xi,x,y) & \longmapsto & (x,y,\xi x,\xi y)\end{array}$$

is an isomorphism of varieties. As $\mathbf{Y}/G\simeq\mathbf{A}^1(\mathbb{F})$ (see Theorem 2.2.2), we conclude that $\mathbf{Z}_0/G\simeq\mu_{q+1}\times\mathbf{A}^1$, with the action of $(\xi,\xi')\in\mu_{q+1}\times\mu_{q+1}$ on $(\zeta,x)\in\mu_{q+1}\times\mathbf{A}^1$ being as follows:

$$(\xi,\xi')\cdot(\zeta,x)=(\xi^{-1}\xi'\zeta,x).$$

Therefore, by Theorems A.2.1(c) and (f) and A.2.6(c), we have

(β) $$H^*_c(\mathbf{Z}_0/G)=[K[\mu^{(2)}_{q+1}]]_{\mu_{q+1}\times\mu_{q+1}}.$$

We now study the quotient $\mathbf{Z}_{\neq 0}/G$. To this end, consider the variety

$$\mathbf{V}=\{(u,a,b)\in(\mathbf{A}^1(\mathbb{F})\setminus\{0\})\times\mathbf{A}^2(\mathbb{F})\mid u^{q+1}-ab=1\}$$

together with the morphism

$$\begin{array}{cccc}\nu: & \mathbf{Z}_{\neq 0} & \longrightarrow & \mathbf{V}\\ & (x,y,z,t) & \longmapsto & (xt-yz,\,xt^q-yz^q,\,x^qt-y^qz).\end{array}$$

A painstaking but easy calculation shows that ν does indeed take values in $\mathbf{V}$ and that $\nu(g\cdot(x,y,z,t)) = \nu(x,y,z,t)$ if $(x,y,z,t) \in \mathbf{Z}_{\neq 0}$ and $g \in G$. To show that ν induces an isomorphism $\bar{\nu}\colon \mathbf{Z}_{\neq 0}/G \simeq \mathbf{V}$, all that remains it to show the following four properties (see Proposition 2.2.1):

(a) ν is surjective.
(b) $\nu(x,y,z,t) = \nu(x',y',z',t')$ if and only if there exists $g \in G$ such that $(x',y') = g\cdot(x,y)$ and $(z',t') = g\cdot(z,t)$.
(c) $\mathbf{V}$ and $\mathbf{Z}_{\neq 0}$ are smooth varieties.
(d) If $m \in \mathbf{Z}_{\neq 0}$, then the differential $d_m\nu$ is surjective.

Let us first prove (a) and (b). Taking into account Proposition 2.1.2, it is enough to show that, if $(u,a,b) \in \mathbf{V}$, then

$$|\nu^{-1}(u,a,b)| = |G|.$$

Now, $\nu^{-1}(u,a,b)$ is the set of quadruples (x,y,z,t) satisfying

$$\begin{cases} xy^q - yx^q = 1 & (1) \\ zt^q - tz^q = 1 & (2) \\ xt - yz = u & (3) \\ xt^q - yz^q = a & (4) \\ x^q t - y^q z = b. & (5) \end{cases}$$

The two equations (3) and (5), viewed as equations in t and z, form a system of linear equations which, by (1), has determinant -1. As a consequence, $(x,y,z,t) \in \nu^{-1}(u,a,b)$ if and only if (x,y,z,t) satisfy the system

$$\begin{cases} xy^q - yx^q = 1 & (1) \\ zt^q - tz^q = 1 & (2) \\ z = ux^q - bx & (3') \\ xt^q - yz^q = a & (4) \\ t = uy^q - by. & (5') \end{cases}$$

If we substitute z and t into (4), we obtain the equivalent system of equations (because $u \neq 0$):

$$\begin{cases} xy^q - yx^q = 1 & (1) \\ zt^q - tz^q = 1 & (2) \\ z = ux^q - bx & (3') \\ xy^{q^2} - yx^{q^2} = \dfrac{a+b^q}{u^q} & (4') \\ t = uy^q - by. & (5') \end{cases}$$

On the other hand, it is easy to verify that (1), (3′), (4′) and (5′) imply (2). The system is therefore equivalent to

$$\begin{cases} xy^q - yx^q = 1 & (1) \\ z = ux^q - bx & (3') \\ xy^{q^2} - yx^{q^2} = \dfrac{a+b^q}{u^q} & (4') \\ t = uy^q - by. & (5') \end{cases}$$

But, by Theorem 2.2.2, the number of couples (x,y) satisfying (1) to (4′) is equal to $|G|$. Because a couple (z,t) is determined by (u,a,b) and (x,y), we indeed have $|v^{-1}(u,a,b)| = |G|$, as expected.

We now turn to (c) and (d). Let $m = (x_0, y_0, z_0, t_0) \in \mathbf{Z}_{\neq 0}$. The tangent space $\mathscr{T}$ to $\mathbf{Z}_{\neq 0}$ at m may be identified with

$$\mathscr{T} = \{(x,y,z,t) \in \mathbb{F}^4 \mid y_0^q x - x_0^q y = 0 \text{ and } t_0^q z - z_0^q t = 0\}.$$

Set $(u_0, a_0, b_0) = v(m)$. The tangent space $\mathscr{T}'$ of the variety $\mathbf{V}$ at $v(m)$ may be identified with

$$\mathscr{T}' = \{(u,a,b) \in \mathbb{F}^3 \mid u_0^q u - b_0 a - a_0 b = 0\}.$$

Using these identifications it is easy to verify that the morphism $d_m v \colon \mathscr{T} \to \mathscr{T}'$ takes the following form:

$$d_m v(x,y,z,t) = (t_0 x + x_0 t - y_0 z - z_0 y, t_0^q x - z_0^q y, x_0^q t - y_0^q z).$$

It is enough to show that $d_m v$ is injective. Now, if $d_m v(x,y,z,t) = (0,0,0)$, then in particular

$$\begin{cases} y_0^q x - x_0^q y = 0 \\ t_0^q z - z_0^q t = 0 \\ t_0^q x - z_0^q y = 0 \\ x_0^q t - y_0^q z = 0. \end{cases}$$

We indeed obtain $x = y = z = t = 0$ because, as $m \in \mathbf{Z}_{\neq 0}$, we have $x_0^q t_0^q - y_0^q z_0^q = (x_0 t_0 - y_0 z_0)^q \neq 0$.

As (a), (b), (c) and (d) hold, the morphism $v \colon \mathbf{Z}_{\neq 0} \to \mathbf{V}$ induces an isomorphism $\bar{v} \colon \mathbf{Z}_{\neq 0}/G \xrightarrow{\sim} \mathbf{V}$ and, under this isomorphism, the action of $(\xi, \xi') \in \mu_{q+1} \times \mu_{q+1}$ is as follows:

$$(\xi, \xi') \cdot (u,a,b) = (\xi\xi' u, \xi\xi'^{-1} a, \xi^{-1}\xi' b).$$

On the other hand, the torus $\mathbb{F}^\times$ acts on $\mathbf{V}$ by the formula:

$$\lambda \cdot (u,a,b) = (u, \lambda a, \lambda^{-1} b),$$

and this action commutes with that of $\mu_{q+1} \times \mu_{q+1}$. As a consequence, by Theorem A.2.6(e), we have

$$H_c^*(\mathbf{Z}_{\neq 0}/G)_{\mu_{q+1}\times\mu_{q+1}} = H_c^*(\mathbf{V})_{\mu_{q+1}\times\mu_{q+1}} = H_c^*(\mathbf{V}^{\mathbb{F}^\times})_{\mu_{q+1}\times\mu_{q+1}}.$$

Now $\mathbf{V}^{\mathbb{F}^\times} = \mu_{q+1} \times \{0\} \times \{0\}$. Therefore

$$(\gamma) \qquad H_c^*(\mathbf{Z}_{\neq 0}/G)_{\mu_{q+1}\times\mu_{q+1}} = [K[\mu_{q+1}^{(1)}]]_{\mu_{q+1}\times\mu_{q+1}}.$$

Equation 4.2.2 follows immediately from (α), (β) and (γ). □

In order to prove the Mackey formula, it will be sufficient to deduce some algebraic consequences from the previous work, which was of a geometric nature.

Proof (of the Mackey formula 4.2.1). We may and will suppose that θ and η are linear characters of μ_{q+1}. By Proposition 4.1.3, we have

$$\langle \mathrm{R}'(\theta), \mathrm{R}'(\eta)\rangle_G = \langle 1_G, \mathrm{R}'(\theta)\cdot \mathrm{R}'(\eta)\rangle_G.$$

As a consequence,

$$\langle \mathrm{R}'(\theta), \mathrm{R}'(\eta)\rangle_G = \dim_K \Big((H_c^*(\mathbf{Y}) \otimes_{K\mu_{q+1}} K_\theta) \otimes_K (H_c^*(\mathbf{Y}) \otimes_{K\mu_{q+1}} K_\theta) \Big)^G.$$

By Theorem A.2.6(b) and (c), we have

$$\langle \mathrm{R}'(\theta), \mathrm{R}'(\eta)\rangle_G = \dim_K H_c^*((\mathbf{Y}\times\mathbf{Y})/G) \otimes_{K(\mu_{q+1}\times\mu_{q+1})} K_{\theta\boxtimes\eta},$$

where $\theta\boxtimes\eta : \mu_{q+1}\times\mu_{q+1} \to K^\times$, $(\xi,\xi') \mapsto \theta(\xi)\eta(\xi')$. It therefore follows from 4.2.2 that

$$\begin{aligned}\langle \mathrm{R}'(\theta), \mathrm{R}'(\eta)\rangle_G &= \dim_K K[\mu_{q+1}^{(1)}] \otimes_{K(\mu_{q+1}\times\mu_{q+1})} K_{\theta\boxtimes\eta} \\ &\quad + \dim_K K[\mu_{q+1}^{(2)}] \otimes_{K(\mu_{q+1}\times\mu_{q+1})} K_{\theta\boxtimes\eta}.\end{aligned}$$

Set $\mu^{(1)} = \{(\xi,\xi^{-1}) \mid \xi \in \mu_{q+1}\}$ and $\mu^{(2)} = \{(\xi,\xi) \mid \xi \in \mu_{q+1}\}$. Then

$$K[\mu_{q+1}^{(i)}] = \mathrm{Ind}_{\mu^{(i)}}^{\mu_{q+1}\times\mu_{q+1}} 1_{\mu^{(i)}}.$$

Therefore, by Frobenius reciprocity,

$$\langle \mathrm{R}'(\theta), \mathrm{R}'(\eta)\rangle_G = \langle 1_{\mu^{(1)}}, \mathrm{Res}_{\mu^{(1)}}^{\mu_{q+1}\times\mu_{q+1}} \theta\boxtimes\eta\rangle_{\mu^{(1)}} + \langle 1_{\mu^{(2)}}, \mathrm{Res}_{\mu^{(2)}}^{\mu_{q+1}\times\mu_{q+1}} \theta\boxtimes\eta\rangle_{\mu^{(2)}},$$

and the result follows. □

4.3. Parametrisation of Irr G

The Mackey formula 4.2.1 together with 4.1.4 show that, if we denote by θ_0 the unique character of μ_{q+1} of order 2:

- $\mathsf{R}'(1) = -1_G + \mathsf{St}_G$.
- $\mathsf{R}'(\theta) = \mathsf{R}'(\theta^{-1}) \in \operatorname{Irr} G$ if $\theta \in \mu_{q+1}^{\wedge}$ is such that $\theta^2 \neq 1$.
- $\mathsf{R}'(\theta_0) = \mathsf{R}'_+(\theta_0) + \mathsf{R}'_-(\theta_0)$, where $\mathsf{R}'_\pm(\theta_0) \in \operatorname{Irr} G$ and $\mathsf{R}'_+(\theta_0) \neq \mathsf{R}'_-(\theta_0)$.
- If $\theta^2 \neq 1$, $\eta^2 \neq 1$ and $\theta \notin \{\eta, \eta^{-1}\}$, then $\mathsf{R}'(\theta) \neq \mathsf{R}'(\eta)$.

We obtain a formula analogous to 3.2.13:

$$-\mathsf{R}'(\mathrm{reg}_{\mu_{q+1}}) = \underbrace{-1_G + \mathsf{St}_G}_{\mathsf{R}'(1)} + \underbrace{\mathsf{R}'_+(\theta_0) + \mathsf{R}'_-(\theta_0)}_{\mathsf{R}'(\theta_0)} + \sum_{\theta^2 \neq 1} \mathsf{R}'(\theta). \tag{4.3.1}$$

We have therefore obtained $(q+3)/2$ cuspidal characters. In Chapter 3, we obtained $(q+5)/2$ non-cuspidal irreducible characters. As $|\operatorname{Irr} G| = q+4$, we conclude that all the cuspidal characters of G have been obtained in the cohomology of $\mathbf{Y}$. Therefore we have

$$\begin{aligned} \operatorname{Irr} G = {} & \{1_G, \mathsf{St}_G\} \,\dot\cup\, \{\mathsf{R}_+(\alpha_0), \mathsf{R}_-(\alpha_0), \mathsf{R}'_+(\theta_0), \mathsf{R}'_-(\theta_0)\} \\ & \dot\cup\, \{\mathsf{R}(\alpha) \mid \alpha \in \mu_{q-1}^{\wedge}, \alpha^2 \neq 1\} \,\dot\cup\, \{\mathsf{R}'(\theta) \mid \theta \in \mu_{q+1}^{\wedge}, \theta^2 \neq 1\}. \end{aligned} \tag{4.3.2}$$

If we denote by $d_\pm$ the degree of $\mathsf{R}'(\theta_0)^\pm$, we have:

$$|G| = 1^2 + q^2 + \frac{q-3}{2}(q+1)^2 + 2\Big(\frac{q+1}{2}\Big)^2 + \frac{q-1}{2}(q-1)^2 + d_+^2 + d_-^2.$$

Hence

$$d_+ + d_- = q-1 \quad \text{and} \quad d_+^2 + d_-^2 = \frac{(q-1)^2}{2},$$

which implies that

$$d_+ = d_- = \frac{q-1}{2}. \tag{4.3.3}$$

In the Exercises 4.3 and 4.4 we will give two other (more conceptual) proofs of this last equality: these exercises imply that $\mathsf{R}'_+(\theta_0)$ and $\mathsf{R}'_-(\theta_0)$ are conjugate under $\widetilde{G} = \mathrm{GL}_2(\mathbb{F}_q)$.

4.4. Action of the Frobenius Endomorphism

We will now study the eigenvalues of the action of F on $H_c^i(\mathbf{Y})$. By Theorem A.2.7(b), we have

(4.4.1) $$F = q \text{ on } H_c^2(\mathbf{Y}).$$

Taking this into account, in what follows we will only be interested in the eigenvalues of the action of F on $H_c^1(\mathbf{Y})$.

If θ is a linear character of μ_{q+1}, the KG-modules $H_c^i(\mathbf{Y}) \otimes_{K\mu_{q+1}} K_\theta$ and $H_c^i(\mathbf{Y})e_\theta$ are canonically isomorphic. Recall that

$$e_\theta = \frac{1}{q+1} \sum_{\xi \in \mu_{q+1}} \theta(\xi^{-1})\, \xi \in K\mu_{q+1}.$$

The commutation relations between F and μ_{q+1} (and the fact that F is an automorphism of $H_c^i(\mathbf{Y})$ by Theorem A.2.7(c)) show that

(4.4.2) $$F(H_c^i(\mathbf{Y})e_\theta) = H_c^i(\mathbf{Y})e_{\theta^{-1}}.$$

In particular, F stabilises $H_c^1(\mathbf{Y})e_1$ and $H_c^1(\mathbf{Y})e_{\theta_0}$.

4.4.1. *Action on* $H_c^1(\mathbf{Y})e_1$

The KG-module $H_c^1(\mathbf{Y})e_1 \simeq H_c^1(\mathbf{Y}/\mu_{q+1})$ is irreducible and therefore, by Schur's lemma and the fact that the action of F commutes with that of G, F acts on $H_c^1(\mathbf{Y})e_1$ by multiplication by a scalar ρ_1. To calculate ρ_1, it is enough to calculate the action of the Frobenius endomorphism F on $H_c^1(\mathbf{Y}/\mu_{q+1}) = H_c^1(\mathbf{P}^1(\mathbb{F}) \setminus \mathbf{P}^1(\mathbb{F}_q))$ (see Theorem 2.2.4). Now, the Lefschetz fixed-point theorem (see Theorem A.2.7(a)) shows that

$$0 = |(\mathbf{P}^1(\mathbb{F}) \setminus \mathbf{P}^1(\mathbb{F}_q))^F| = q - \rho_1 \dim_K H_c^1(\mathbf{Y}/\mu_{q+1}) = q(1-\rho_1).$$

Hence

(4.4.3) $$\rho_1 = 1.$$

4.4.2. Action on $H_c^1(\mathbf{Y})e_{\theta_0}$

To simplify notation set $V_{\theta_0} = H_c^1(\mathbf{Y})e_{\theta_0}$. Denote by $V_{\theta_0}^{\pm}$ the irreducible subrepresentation of V_{θ_0} with character $R'_{\pm}(\theta_0)$. By Schur's lemma, F acts on $V_{\theta_0}^{\pm}$ by multiplication by a scalar $\rho_{\pm}$. We would like to calculate $\rho_{\pm}$.

Firstly, we have $\mathbf{Y}^F = \varnothing$ and therefore, by the Lefschetz fixed-point formula, we obtain

$$0 = q - q\rho_1 - \frac{(q-1)(\rho_+ + \rho_-)}{2} - \mathrm{Tr}(F, \bigoplus_{\theta^2 \neq 1} H_c^1(\mathbf{Y})e_\theta).$$

But

$$\mathrm{Tr}(F, \bigoplus_{\theta^2 \neq 1} H_c^1(\mathbf{Y})e_\theta) = 0$$

by 4.4.2. Hence

(4.4.4) $$\rho_- = -\rho_+.$$

To explicitly calculate ρ_+ and ρ_-, we will study the action of F^2. As F^2 stabilises $H_c^1(\mathbf{Y})e_\theta$, it follows from Schur's lemma that F^2 acts on $H_c^1(\mathbf{Y})e_\theta$ by multiplication by a scalar λ_θ (in fact, if $\theta = \theta_0$, then this follows in fact from 4.4.4 as $\rho_+^2 = \rho_-^2$).

Theorem 4.4.5. *Let $\theta \in (\mu_{q+1})^\wedge$. Then*

$$\lambda_\theta = \begin{cases} 1 & \text{if } \theta = 1, \\ -\theta(-1)q & \text{if } \theta \neq 1. \end{cases}$$

Proof. The equality $\lambda_1 = 1$ follows from 4.4.3. The Lefschetz fixed-point theorem shows that

$$|\mathbf{Y}^{\xi F^2}| = q^2 - q\lambda_1 - \sum_{\theta \neq 1} (q-1)\theta(\xi)\lambda_\theta$$

for all $\xi \in \mu_{q+1}$. As a consequence, as $\lambda_1 = 1$, we have

$$|\mathbf{Y}^{\xi F^2}| = q^2 - 1 - (q-1) \sum_{\theta \in \mu_{q+1}^\wedge} \theta(\xi)\lambda_\theta.$$

It then follows from Theorem 2.3.2 that

$(\mathscr{E}_\xi)$ $$\sum_{\theta \in \mu_{q+1}^\wedge} \theta(\xi)\lambda_\theta = (q+1) - \frac{|\mathbf{Y}^{\xi F^2}|}{q-1} = \begin{cases} 1 - q^2 & \text{if } \xi = -1, \\ q+1 & \text{if } \xi \neq -1. \end{cases}$$

The family of eigenvalues $(\lambda_\theta)_{\theta \in (\mu_{q+1})^\wedge}$ is therefore a solution of the system of linear equations $(\mathscr{E}_\xi)_{\xi \in \mu_{q+1}}$. The determinant of this system is invertible (being the determinant of the character table of a cyclic group) and therefore it admits a unique solution. It remains only to verify that the solutions given in the corollary are valid, which is routine. □

Corollary 4.4.6. *We have* $\rho_\pm = \pm\sqrt{-\theta_0(-1)q}$.

4.4.3. *Action on* $H^1_c(\mathbf{Y})e_\theta \oplus H^1_c(\mathbf{Y})e_{\theta^{-1}}$

Let $\theta \in \mu_{q+1}^\wedge$ be such that $\theta^2 \neq 1$. Then F stabilises $H^1_c(\mathbf{Y})e_\theta \oplus H^1_c(\mathbf{Y})e_{\theta^{-1}}$ and F^2 acts as multiplication by $-\theta(-1)q$. Therefore F has two eigenvalues, $\sqrt{-\theta(-1)q}$ and $-\sqrt{-\theta(-1)q}$, each one having multiplicity $q-1$ (because, as $F(H^1_c(\mathbf{Y})e_\theta) = H^1_c(\mathbf{Y})e_{\theta^{-1}}$, the trace of F on the direct sum is zero).

Exercises

4.1* (Lusztig). In this exercise we show that

$$(*)\qquad \mathsf{R}'_+(\theta_0)^* = \begin{cases} \mathsf{R}'_+(\theta_0) & \text{si } q \equiv 1 \mod 4,\\ \mathsf{R}'_-(\theta_0) & \text{si } q \equiv 3 \mod 4.\end{cases}$$

We will use Poincaré duality, which is only possible using the (smooth) compactification $\overline{\mathbf{Y}}$ of $\mathbf{Y}$. Denote by $\langle,\rangle\colon H^1_c(\overline{\mathbf{Y}}) \times H^1_c(\overline{\mathbf{Y}}) \to H^2_c(\overline{\mathbf{Y}})$ the perfect pairing A.2.4.

(a) Show that, if $\theta \neq 1$, then the KG-modules $H^1_c(\mathbf{Y})e_\theta$ and $H^1_c(\overline{\mathbf{Y}})e_\theta$ are isomorphic (***Hint:*** Use the open-closed exact sequence of Theorem A.2.1(d)).

To simplify notation set $V_{\theta_0} = H^1_c(\overline{\mathbf{Y}})e_{\theta_0} \simeq H^1_c(\mathbf{Y})e_{\theta_0}$ and denote by $V^\pm_{\theta_0}$ the irreducible submodule of V_{θ_0} with character $\mathsf{R}'_\pm(\theta_0)$.

(b) Show that $\langle,\rangle$ induces a perfect KG-equivariant pairing between $H^1_c(\overline{\mathbf{Y}})e_\theta$ and $H^1_c(\overline{\mathbf{Y}})e_{\theta^{-1}}$. Deduce that $\mathsf{R}'(\theta)^* = \mathsf{R}'(\theta)$.

We will denote by $\langle,\rangle_0$ the perfect pairing on V_{θ_0} obtained by restriction from $\langle,\rangle$.

(c) Show that $\theta_0(-1) = \begin{cases} 1 & \text{if } q \equiv 3 \mod 4,\\ -1 & \text{if } q \equiv 1 \mod 4.\end{cases}$

(d) Suppose that $q \equiv 1 \mod 4$. Show that $V^+_{\theta_0}$ is orthogonal to $V^-_{\theta_0}$. (***Hint:*** Use the F-equivariance of $\langle,\rangle_0$ and the knowledge of the eigenvalues of F). Deduced that $\langle,\rangle_0$ induces an isomorphism of KG-modules $V^+_{\theta_0} \simeq (V^+_{\theta_0})^*$ (and therefore that $\mathsf{R}'_+(\theta_0)^* = \mathsf{R}'_+(\theta_0)$).

(e) Suppose now that $q \equiv 3 \mod 4$. Show that the restriction of $\langle,\rangle_{\theta_0}$ to $V_{\theta_0}^+$ is zero (***Hint:*** Use the F-equivariance of $\langle,\rangle_0$ and the knowledge of the eigenvalues of F). Deduce that $\langle,\rangle_0$ induces an isomorphism of KG-modules $V_{\theta_0}^+ \simeq (V_{\theta_0}^-)^*$ (and therefore that $\mathrm{R}'_+(\theta_0)^* = \mathrm{R}'_-(\theta_0)$).

REMARK – Part (c) shows that $-\theta_0(-1) = \alpha_0(-1)$ (see the remark following Exercise 3.3). □

4.2. Show that G has four irreducible characters of odd degree.

4.3. Denote by $A = \widetilde{G}/G \cdot \mathrm{Z}(\widetilde{G})$. Recall that A is of order 2 (see Exercise 3.1). The group A acts on the conjugacy classes and on the irreducible characters of G. In this exercise, we will give another proof of 4.3.3.

(a) Show that A stabilises the q semi-simple conjugacy classes of G and permutes the other four without fixed points.
(b) Show that A stabilises the q characters 1_G, St_G, $\mathrm{R}(\alpha)$ $(\alpha \in (\mu_{q-1})^\wedge, \alpha^2 \neq 1)$ and $\mathrm{R}'(\theta)$ $(\theta \in (\mu_{q+1})^\wedge, \theta^2 \neq 1)$. (***Hint:*** Use the group $\mathscr{G}$.)
(c) Deduce that A permutes the four characters $\mathrm{R}_\pm(\alpha_0)$ and $\mathrm{R}'_\pm(\theta_0)$ without fixed points. (***Hint:*** Apply some theorem of Brauer [Isa, Theorem 6.32].)
(d) Deduce that $\mathrm{R}'_+(\theta_0)$ and $\mathrm{R}_-(\theta_0)$ are conjugate under $\widetilde{G}$ and that $d_+ = d_- = (q-1)/2$.

4.4* In this exercise, we define a Deligne-Lusztig induction for $\widetilde{G}$ and deduce that the characters $\mathrm{R}'_+(\theta_0)$ and $\mathrm{R}'_-(\theta_0)$ are conjugate under $\widetilde{G}$ (which gives a new proof of 4.3.3). Set

$$\widetilde{\mathbf{Y}} = \{(x,y) \in \mathbf{A}^2(\mathbb{F}) \mid (xy^q - yx^q)^{q-1} = 1\}.$$

(a) Show that $\widetilde{G} \times \mathbb{F}_{q^2}^\times$ acts naturally on $\widetilde{\mathbf{Y}}$. We let $\mathbb{F}_{q^2}^\times$ act on the right.
(b) Show that $\widetilde{\mathbf{Y}} = \widetilde{G} \times^G \mathbf{Y} = \mathbf{Y} \times_{\mu_{q+1}} \mathbb{F}_{q^2}^\times$. Here, $\mathbf{A} \times^\Gamma \mathbf{B}$ denotes the quotient of $\mathbf{A} \times \mathbf{B}$ by the diagonal action of Γ.

Set $\widetilde{\mathrm{R}}' : \mathscr{K}_0(K\mathbb{F}_{q^2}^\times) \to \mathscr{K}_0(K\widetilde{G}), [M]_{\mathbb{F}_{q^2}^\times} \mapsto -\sum_{i \geqslant 0}(-1)^i [H_c^i(\widetilde{\mathbf{Y}}) \otimes_{K\mathbb{F}_{q^2}^\times} M]_{\widetilde{G}}$.

(c) Deduce from (b) that $\widetilde{\mathrm{R}}' \circ \mathrm{Res}_{\mu_{q+1}}^{\mathbb{F}_{q^2}^\times} = \mathrm{Res}_G^{\widetilde{G}} \circ \widetilde{\mathrm{R}}'$.
(d*) Show that

$$\langle \widetilde{\mathrm{R}}'(\theta), \widetilde{\mathrm{R}}'(\eta) \rangle_{\widetilde{G}} = \langle \theta, \eta \rangle_{\mathbb{F}_{q^2}^\times} + \langle \theta, {}^F\eta \rangle_{\mathbb{F}_{q^2}^\times}.$$

(***Hint:*** Mimic the proof of the Mackey formula 4.2.1 for the group G.)
(e) Let $\tilde{\theta}_0$ be an extension of θ_0 to $\mathbb{F}_{q^2}^\times$. Show that ${}^F\tilde{\theta}_0 \neq \tilde{\theta}_0$.
(f) Deduce that $\widetilde{\mathrm{R}}'(\tilde{\theta}_0)$ is irreducible, that $\mathrm{R}'(\theta_0) = \mathrm{Res}_G^{\widetilde{G}} \widetilde{\mathrm{R}}'(\tilde{\theta}_0)$, and that $\mathrm{R}'_+(\theta_0)$ and $\mathrm{R}'_-(\theta_0)$ are conjugate under $\widetilde{G}$.
(g) Show that $d_+ = d_- = (q-1)/2$.

4.5. Let m be a non-zero natural number. Show that

$$|\overline{\mathbf{Y}}^{F^m}| = |\mathbf{Y}^{F^m}| + q + 1$$

and

$$|\overline{\mathbf{Y}}^{F^m}| = \begin{cases} q^m + 1 & \text{if } m \text{ is odd,} \\ q^m + 1 - q(q-1)q^{m/2} & \text{if } m \equiv 0 \mod 4, \\ q^m + 1 - (q-1)q^{m/2} & \text{if } m \equiv 2 \mod 4. \end{cases}$$

Verify that $|\mathbf{Y}^{F^m}|$ is divisible by $|G|$.

4.6. Let $\tilde{\mathbf{V}} = \{(u,a,b) \in \mathbf{A}^3(\mathbb{F}) \mid u^{q+1} - ab = 1\}$ and denote by $\tilde{v}\colon \mathbf{Z} \to \tilde{\mathbf{V}}$ the morphism defined by

$$\tilde{v}(x,y,z,t) = (xt - yz, xt^q - yz^q, x^q t - y^q z).$$

We also use the notation introduced in the proof of the Mackey formula ($\mathbf{Z}$, $\mathbf{V}$, v,...).

(a) Show that $\mathbf{V}$ is an open dense subset of $\tilde{\mathbf{V}}$ and that $\tilde{v}$ is a well-defined extension of the morphism $v\colon \mathbf{Z}_{\neq 0} \to \mathbf{V}$.
(b) Show that $\tilde{\mathbf{V}}$ and $\mathbf{Z}$ are smooth.
(c) Show that $\tilde{v}$ is constant on G-orbits.
(d) Nevertheless, show that v does not induce an isomorphism between $\mathbf{Z}/G$ and $\tilde{\mathbf{V}}$.

Chapter 5
The Character Table

Having parametrised the irreducible characters of the group G in the last chapter, it is natural to turn to the question of determining their values on the elements of G. For this, an important step is the calculation of the characters of the bimodules $K[G/U]$ and $H_c^i(\mathbf{Y})$. For $K[G/U]$, a little elementary linear algebra is sufficient. For $H_c^i(\mathbf{Y})$ it is necessary to invoke certain results from Appendix A.

These calculations allow us to calculate the majority of the irreducible characters of G, but it does not allow us to determine the values of $\mathsf{R}_\pm(\alpha_0)$ and $\mathsf{R}'_\pm(\theta_0)$. In the latter case, we study their restriction to U and utilise certain elementary arithmetic results (on *Gauss sums*).

5.1. Characters of Bimodules

Denote by

$$\begin{array}{rccc} \mathfrak{Tr}\colon & G \times \mu_{q-1} & \longrightarrow & K \\ & (g,a) & \longmapsto & \mathsf{Tr}((g,\mathbf{d}(a)),K[G/U]) \end{array}$$

and

$$\begin{array}{rccc} \mathfrak{Tr}'\colon & G \times \mu_{q+1} & \longrightarrow & K \\ & (g,\xi) & \longmapsto & -\mathsf{Tr}^*_{\mathbf{Y}}(g,\xi). \end{array}$$

The goal of this section is to calculate $\mathfrak{Tr}$ and $\mathfrak{Tr}'$.

5.1.1. *Calculation of* $\mathfrak{Tr}$

The bimodule $K[G/U]$ is a permutation bimodule. Therefore, if $(g,t) \in G \times T$, then

$$\mathsf{Tr}((g,t),K[G/U]) = |\{xU \in G/U \mid x^{-1}gx \in t^{-1}U\}|.$$

C. Bonnafé, *Representations of* $\mathrm{SL}_2(\mathbb{F}_q)$, Algebra and Applications 13,
DOI 10.1007/978-0-85729-157-8_5,

From this we easily deduce the values of $\mathfrak{Tr}(g,b)$, which are given in Table 5.1.

Table 5.1 Values of $\mathfrak{Tr}$

	$\varepsilon \mathrm{I}_2$ $\varepsilon \in \{\pm 1\}$	$\mathbf{d}(a)$ $a \in \mu_{q-1} \setminus \{\pm 1\}$	$\mathbf{d}'(\xi)$ $\xi \in \mu_{q+1} \setminus \{\pm 1\}$	$\begin{pmatrix} \varepsilon & x \\ 0 & \varepsilon \end{pmatrix}$ $\varepsilon \in \{\pm 1\}, x \in \mathbb{F}_q^\times$
b	$(q^2-1)\delta_{b=\varepsilon}$	$(q-1)\delta_{b \in \{a,a^{-1}\}}$	0	$(q-1)\delta_{b=\varepsilon}$

In this and following tables, if P is a statement, then δ_P has value 1 if P is true and is 0 otherwise.

5.1.2. *Calculation of* $\mathfrak{Tr}'$

Let $(g,\xi) \in G \times \mu_{q+1}$. We write $g = tu = ut$, where t is of order prime to p and u is of order a power of p (*Jordan decomposition*). By Theorem A.2.6(d), we have

$$\mathfrak{Tr}'(g,\xi) = -\operatorname{Tr}^*_{\mathbf{Y}^{(t,\xi)}}(u).$$

Hence we must calculate the fixed points of (t,ξ) on $\mathbf{Y}$: this is easy and the result is given in the following lemma.

Lemma 5.1.1. *We have*

$$\mathbf{Y}^{(t,\xi)} = \begin{cases} \mathbf{Y} & \text{if } \xi^2 = 1 \text{ and } t = \xi \mathrm{I}_2, \\ \mu_{q+1} v_\xi & \text{if } \xi^2 \neq 1 \text{ and } t \text{ is conjugate to } \mathbf{d}'(\xi), \\ \varnothing & \text{otherwise.} \end{cases}$$

Here, $v_\xi \in \mathbf{Y}$ *satisfies* $t \cdot v_\xi = \xi^{-1} v_\xi$.

Corollary 5.1.2. *If* ξ *is not an eigenvalue of* g*, then* $\mathfrak{Tr}'(g,\xi) = 0$.

Recall that u_+ and u_- are representatives (in U) of the conjugacy classes of non-trivial unipotent elements of G.

Lemma 5.1.3. $\mathfrak{Tr}'(u_+,1) = \mathfrak{Tr}'(u_-,1) = -(q+1)$.

Proof. Set $\lambda = \mathfrak{Tr}'(u_+,1)$. As u_+ and u_- are conjugate under $\mathrm{GL}_2(\mathbb{F}_q)$, the elements $(u_+,1)$ and $(u_-,1)$ of $\mathscr{G}$ (see §2.5.1 for the definition of $\mathscr{G}$) are conjugate under $\mathscr{G}$. Therefore $\lambda = \mathfrak{Tr}'(u_-,1)$. Now, taking into account 5.1.2, we have

$$\mathsf{R}'(\mathrm{reg}_{\mu_{q+1}})(g) = \begin{cases} q^2-1 & \text{if } g = 1, \\ \lambda & \text{if } g \text{ is conjugate to } u_+ \text{ or } u_-, \\ 0 & \text{otherwise.} \end{cases}$$

Moreover, by Exercise 3.2,

$$\mathsf{R}(1)(g) = \begin{cases} q+1 & \text{if } g = 1, \\ 1 & \text{if } g \text{ is conjugate to } u_+ \text{ or } u_-. \end{cases}$$

But, by Theorem 4.1.7, we have $\langle \mathsf{R}(1), \mathsf{R}'(\mathrm{reg}_{\mu_{q+1}})\rangle_G = 0$. In other words,

$$(q+1)(q^2-1) + (q^2-1)\lambda = 0.$$

The result now follows easily. □

Thus we have obtained all necessary information to easily determine the values of $\mathfrak{Tr}'(g, \xi')$, which are given in Table 5.2.

Table 5.2 Values of $\mathfrak{Tr}'$

	εI_2 $\varepsilon \in \{\pm 1\}$	$\mathbf{d}(a)$ $a \in \mu_{q-1} \setminus \{\pm 1\}$	$\mathbf{d}'(\xi)$ $\xi \in \mu_{q+1} \setminus \{\pm 1\}$	$\begin{pmatrix} \varepsilon & x \\ 0 & \varepsilon \end{pmatrix}$ $\varepsilon \in \{\pm 1\}, x \in \mathbb{F}_q^\times$
ξ'	$(q^2-1)\delta_{\xi'=\varepsilon}$	0	$-(q+1)\delta_{\xi' \in \{\xi, \xi^{-1}\}}$	$-(q+1)\delta_{\xi'=\varepsilon}$

5.1.3. *The Characters* $\mathsf{R}(\alpha)$ *and* $\mathsf{R}'(\theta)$

Fix two linear characters α and θ of μ_{q-1} and μ_{q+1} respectively. Then the values of $\mathsf{R}(\alpha)$ and $\mathsf{R}'(\theta)$ are given in Table 5.3 (as follows, using 3.1.6 and Tables 5.1 and 5.2).

5.2. Restriction to U

The information contained in Table 5.3 gives much of the character table of G. The only characters which remain to be determined are $\mathsf{R}_\pm(\alpha_0)$ and $\mathsf{R}'_\pm(\theta_0)$. For this, we will determine their restriction to U.

Table 5.3 Values of the characters $\mathsf{R}(\alpha)$ and $\mathsf{R}'(\theta)$

	εI_2 $\varepsilon \in \{\pm 1\}$	$\mathbf{d}(a)$ $a \in \mu_{q-1} \setminus \{\pm 1\}$	$\mathbf{d}'(\xi)$ $\xi \in \mu_{q+1} \setminus \{\pm 1\}$	$\begin{pmatrix} \varepsilon & x \\ 0 & \varepsilon \end{pmatrix}$ $\varepsilon \in \{\pm 1\}, x \in \mathbb{F}_q^\times$
$\mathsf{R}(\alpha)$	$(q+1)\alpha(\varepsilon)$	$\alpha(a)+\alpha(a^{-1})$	0	$\alpha(\varepsilon)$
$\mathsf{R}'(\theta)$	$(q-1)\theta(\varepsilon)$	0	$-\theta(\xi)-\theta(\xi)^{-1}$	$-\theta(\varepsilon)$

5.2.1. *B-Invariant Characters of U*

We fix a non-trivial linear character χ_+ of $\mathbb{F}_q^+$. The morphism

(5.2.1)
$$\begin{array}{rcl} \mathbb{F}_q^+ & \longrightarrow & (\mathbb{F}_q^+)^\wedge \\ z & \longmapsto & (z' \mapsto \chi_+(zz')) \end{array}$$

is an isomorphism of groups. (It is enough to check injectivity, which follows easily from the non-triviality of χ_+.)

Denote by $\mathscr{C}$ the squares in $\mathbb{F}_q^\times$ and let $z_0 \in \mathbb{F}_q^\times \setminus \mathscr{C}$. We set

$$\Upsilon_+ : \begin{array}{rcl} U & \longrightarrow & K \\ \mathbf{u}(z) & \longmapsto & \sum_{c \in \mathscr{C}} \chi_+(cz) \end{array} \qquad \text{and} \qquad \Upsilon_- : \begin{array}{rcl} U & \longrightarrow & K \\ \mathbf{u}(z) & \longmapsto & \sum_{c \in \mathscr{C}} \chi_+(cz_0 z). \end{array}$$

Taking into account Proposition 1.1.2(b), Υ_+ and Υ_- are B-invariant characters of U. Even better, one can easily verify the following lemma (recall that $\widetilde{B}$ is the subgroup of $\widetilde{G} = \mathrm{GL}_2(\mathbb{F}_q)$ formed by upper triangular matrices).

Lemma 5.2.2. *With notation as above, we have:*

(a) $(1_U, \Upsilon_+, \Upsilon_-)$ *is a* $\mathbb{Z}$*-basis of* $\mathscr{K}_0(KU)^B$.
(b) $(1_U, \Upsilon_+ + \Upsilon_-)$ *is a* $\mathbb{Z}$*-basis of* $\mathscr{K}_0(KU)^{\widetilde{B}}$.

Note that

(5.2.3)
$$\Upsilon_+ + \Upsilon_- = \mathrm{reg}_U - 1_U.$$

5.2.2. *Restriction of Characters of G*

Firstly, we have the following.

Proposition 5.2.4. *Let α and θ be linear characters of μ_{q-1} and μ_{q+1} respectively. Then*

$$\mathrm{Res}_U^G \mathrm{R}(\alpha) = \mathrm{reg}_U + 1_U = 2\cdot 1_U + \Upsilon_+ + \Upsilon_-$$

and

$$\mathrm{Res}_U^G \mathrm{R}'(\theta) = \mathrm{reg}_U - 1_U = \Upsilon_+ + \Upsilon_-.$$

Proof. This follows from Table 5.3. □

Corollary 5.2.5. $\mathrm{Res}_U^G \mathrm{St}_G = \mathrm{reg}_U$.

Denote by $\psi_+ = \chi_+ \circ \mathbf{u}^{-1} \colon U \to \mathbb{F}_q^\times$. We have

$$\psi_+ \begin{pmatrix} 1 & z \\ 0 & 1 \end{pmatrix} = \chi_+(z).$$

The previous proposition shows easily the next corollary.

Corollary 5.2.6. $\langle \mathrm{R}(\alpha), \mathrm{Ind}_U^G \psi_+ \rangle_G = \langle \mathrm{R}'(\theta), \mathrm{Ind}_U^G \psi_+ \rangle_G = 1.$

Up until now, the irreducible components of $\mathrm{R}_\pm(\alpha_0)$ and $\mathrm{R}'_\pm(\theta_0)$ of $\mathrm{R}(\alpha_0)$ and $\mathrm{R}'(\theta_0)$ have not yet been singled out. Corollary 5.2.6 gives us a possibility to do so.

Notation. *From now on, we will denote by $\mathrm{R}_+(\alpha_0)$ (respectively $\mathrm{R}_+(\theta_0)$) the unique common irreducible component of $\mathrm{R}(\alpha_0)$ (respectively $\mathrm{R}'(\theta_0)$) and $\mathrm{Ind}_U^G \psi_+$.*

Proposition 5.2.7. *With the above choice, we obtain*

$$\mathrm{Res}_U^G \mathrm{R}_\pm(\alpha_0) = 1_U + \Upsilon_\pm \quad \textit{and} \quad \mathrm{Res}_U^G \mathrm{R}'_\pm(\theta_0) = \Upsilon_\pm.$$

Proof. By construction, $\mathrm{Res}_U^G \mathrm{R}'_+(\theta_0)$ contains ψ_+ and therefore "contains" Υ_+. As $\deg \mathrm{R}'_+(\theta_0) = \deg \Upsilon_+ = (q-1)/2$, we conclude the second equality.

We now turn to the first equality. Set $\Psi_\pm = \mathrm{Res}_U^G \mathrm{R}_\pm(\alpha_0)$ and write $\Psi_\pm = \lambda_\pm 1_U + \mu_\pm \Upsilon_+ + \nu_\pm \Upsilon_\pm$. As $\mathrm{R}_+(\alpha_0)$ and $\mathrm{R}_-(\alpha_0)$ are conjugate under $\widetilde{G}$, Ψ_+ and Ψ_- are conjugate under $\widetilde{B}$. Therefore $\lambda_+ = \lambda_-$. Moreover, $\lambda_+ + \lambda_- = 2$ by Proposition 5.2.4, therefore $\lambda_\pm = 1$. On the other hand, by construction, $\mu_+ \geqslant 1$. As $\deg \Psi_+ = (q+1)/2 = \deg(1_U + \Upsilon_+)$, we conclude that $\Psi_+ = 1_U + \Upsilon_+$. Similarly, $\Psi_- = 1_U + \Upsilon_-$. □

5.2.3. *Values of* $\Upsilon_\pm$

In order to use the information gathered in the previous subsection, we will calculate the values of the characters $\Upsilon_\pm$ at the unipotent elements $u_\pm$. For this we will need to use *Gauss sums* but, luckily, we will only need to consider the simplest case. Let us temporarily define

$$\gamma = \sum_{z\in\mathbb{F}_q^\times} \alpha_0(z)\chi_+(z).$$

Then

(5.2.8) $$\gamma^2 = \alpha_0(-1)q.$$

Proof (of 5.2.8). We have

$$\gamma^2 = \sum_{z,z'\in\mathbb{F}_q^\times} \alpha_0(zz')\chi_+(z+z').$$

Let $z'' = z^{-1}z'$. We obtain

$$\gamma^2 = \sum_{z,z''\in\mathbb{F}_q^\times} \alpha_0(z'')\chi_+(z(1+z'')).$$

Hence

$$\begin{aligned}\gamma^2 &= \sum_{z''\in\mathbb{F}_q^\times} \Big(\alpha_0(z'') \sum_{z\in\mathbb{F}_q^\times} \chi_+(z(1+z''))\Big)\\ &= (q-1)\alpha_0(-1) + \sum_{z''\in\mathbb{F}_q^\times\setminus\{-1\}} \Big(\alpha_0(z'') \sum_{z\in\mathbb{F}_q^\times} \chi_+(z(1+z''))\Big).\end{aligned}$$

But, if $z'' \neq -1$, then

$$\sum_{z\in\mathbb{F}_q^\times} \chi_+(z(1+z'')) = -1 + \sum_{z\in\mathbb{F}_q^+} \chi_+(z(1+z'')) = -1.$$

Finally,

$$\gamma^2 = (q-1)\alpha_0(-1) - \sum_{z''\in\mathbb{F}_q^\times\setminus\{-1\}} \alpha_0(z'') = q\alpha_0(-1) - \sum_{z''\in\mathbb{F}_q^\times} \alpha_0(z'') = q\alpha_0(-1),$$

as expected. □

This allows us to choose a square root of $\alpha_0(-1)q$ in K. We set

$$\sqrt{\alpha_0(-1)q} = \sum_{z\in\mathbb{F}_q^\times} \alpha_0(z)\chi_+(z).$$

Then we have the following lemma.

Lemma 5.2.9. $\Upsilon_+(u_\pm) = \dfrac{-1 \pm \sqrt{\alpha_0(-1)q}}{2}$ *and* $\Upsilon_-(u_\pm) = -\dfrac{1 \pm \sqrt{\alpha_0(-1)q}}{2}$.

Proof. It is enough to note that $\Upsilon_+(u_+) + \Upsilon_-(u_+) = -1$ (see 5.2.3) and that $\Upsilon_+(u_+) - \Upsilon_-(u_+) = \gamma$ (by definition of γ). □

5.3. Character Table

The previous section allows us to calculate the values of the characters $\mathsf{R}_\pm(\alpha_0)$ and $\mathsf{R}'_\pm(\theta_0)$ at non-trivial unipotent elements. To complete the character table, we need only the following results.

Proposition 5.3.1. *Let* $g \in G$, $\alpha \in \mu_{q-1}^\wedge$ *and* $\theta \in \mu_{q+1}^\wedge$. *Then:*

(a) $\mathsf{R}(\alpha)(-g) = \alpha(-1)\mathsf{R}(\alpha)(g)$ *and* $\mathsf{R}'(\theta)(-g) = \theta(-1)\mathsf{R}'(\theta)(g)$.
(b) $\mathsf{R}_\pm(\alpha_0)(-g) = \alpha_0(-1)\mathsf{R}_\pm(\alpha_0)(g)$ *and* $\mathsf{R}'_\pm(\theta_0)(-g) = \theta_0(-1)\mathsf{R}'_\pm(\theta_0)(g)$.
(c) *If g is semi-simple, then* $\mathsf{R}_\pm(\alpha_0)(g) = \frac{1}{2}\mathsf{R}(\alpha_0)(g)$ *and* $\mathsf{R}'_\pm(\theta_0)(g) = \frac{1}{2}\mathsf{R}'(\theta_0)(g)$.

Proof. (a) is clear and (b) follows immediately from (a). The last assertion follows from the fact that $\mathsf{R}_+(\alpha_0)$ and $\mathsf{R}_-(\alpha_0)$ (respectively $\mathsf{R}'_+(\theta_0)$ and $\mathsf{R}'_-(\theta_0)$) are conjugate under $\widetilde{G}$ by Exercise 4.3 or 4.4. □

Having now collected all necessary information, the character table of G is given in Table 5.4. To simplify notation in this table we have set $q_0 = \alpha_0(-1)q$. In other words, q_0 is the unique element of $\{q, -q\}$ such that $q_0 \equiv 1 \bmod 4$. In particular, as has already been pointed out in the remark following Exercise 3.3, the numbers

$$\frac{\pm 1 \pm \sqrt{q_0}}{2}$$

are algebraic integers.

Exercises

5.1. Show that, if $\alpha \in (\mu_{q-1})^\wedge$ and $\theta \in (\mu_{q+1})^\wedge$, then

$$\mathsf{St}_G \cdot \mathsf{R}(\alpha) = \mathsf{Ind}_T^G \alpha \quad \text{et} \quad \mathsf{St}_G \cdot \mathsf{R}'(\theta) = \mathsf{Ind}_{T'}^G \theta.$$

Conclude that

$$\mathsf{St}_G = \frac{1}{2}(\mathsf{Ind}_T^G 1_T - \mathsf{Ind}_{T'}^G 1_{T'})$$

and

$$\mathsf{St}_G \cdot \mathsf{St}_G = \frac{1}{2}(\mathsf{Ind}_T^G 1_T + \mathsf{Ind}_{T'}^G 1_{T'}).$$

Table 5.4 Character table of $G = \mathrm{SL}_2(\mathbb{F}_q)$

g	$\varepsilon \mathrm{I}_2$	$\mathbf{d}(a)$	$\mathbf{d}'(\xi)$	εu_τ
	$\varepsilon \in \{\pm 1\}$	$a \in \mathbb{F}_q^\times \setminus \{\pm 1\}$	$\xi \in \mu_{q+1} \setminus \{\pm 1\}$	$\varepsilon \in \{\pm 1\}, \tau \in \{\pm\}$
$\lvert \mathrm{Cl}_G(g) \rvert$	1	q^2+q	q^2-q	$\dfrac{q^2-1}{2}$
$o(g)$	$o(\varepsilon)$	$o(a)$	$o(\xi)$	$p \cdot o(\varepsilon)$
$C_G(g)$	G	T	T'	$\{\pm \mathrm{I}_2\} \times U$
1_G	1	1	1	1
St_G	q	1	-1	0
$\mathrm{R}(\alpha),\ \alpha^2 \neq 1$	$(q+1)\alpha(\varepsilon)$	$\alpha(a)+\alpha(a)^{-1}$	0	$\alpha(\varepsilon)$
$\mathrm{R}'(\theta),\ \theta^2 \neq 1$	$(q-1)\theta(\varepsilon)$	0	$-\theta(\xi)-\theta(\xi)^{-1}$	$-\theta(\varepsilon)$
$\mathrm{R}_\sigma(\alpha_0),\ \sigma \in \{\pm\}$	$\dfrac{(q+1)\alpha_0(\varepsilon)}{2}$	$\alpha_0(a)$	0	$\alpha_0(\varepsilon)\dfrac{1+\sigma\tau\sqrt{q_0}}{2}$
$\mathrm{R}'_\sigma(\theta_0),\ \sigma \in \{\pm\}$	$\dfrac{(q-1)\theta_0(\varepsilon)}{2}$	0	$-\theta_0(\xi)$	$\theta_0(\varepsilon)\dfrac{-1+\sigma\tau\sqrt{q_0}}{2}$

5.2. Set

$$\Gamma_+ = \mathrm{Ind}_U^G \psi^+.$$

Show that Γ_+ is a multiplicity-free character of G and that $\langle \Gamma_+, \Gamma_+ \rangle_G = q+1$.

REMARK – The character Γ_+ defined in this exercise is called a *Gelfand-Graev character* of G. □

5.3. It was shown in Exercise 3.4 (respectively Exercise 4.1) that $\mathrm{R}_+(\alpha_0)$ (respectively $\mathrm{R}'_+(\theta_0)$) is self-dual if and only if $q \equiv 1 \mod 4$. Rediscover this result by inspecting the character table.

5.4* Calculate the character table of $\widetilde{G} = \mathrm{GL}_2(\mathbb{F}_q)$.

Part III
Modular Representations

In the next five chapters we study the modular representations of G. In Appendix B we recall the necessary facts about modular representations of "abstract" finite groups. For general reductive finite groups, the nature of this study is radically different, depending on whether one is in *unequal characteristic* (where one studies ℓ-modular representations, where ℓ is a prime number different from p) or in *equal characteristic* (where one studies p-modular representations). Here p denotes the characteristic of the "field of definition" of the group.

In unequal characteristic, the geometric methods of Deligne-Lusztig theory remain as powerful as ever; this cohomology theory supplies representations not just over K, but over $\mathcal{O}$ and k as well (see the definition of $\mathcal{O}$ and k in the notation below). It has also been observed that the family of natural numbers d such that ℓ divides $\Phi_d(q)$ and $\Phi_d(q)$ divides the order of the group (Φ_d denotes the d-th cyclotomic polynomial) play a fundamental role. In the case of our group G (whose cardinality is $q\Phi_1(q)\Phi_2(q)$) it will be convenient to distinguish three cases: $\ell = 2$, ℓ is odd and divides $q-1$, ℓ is odd and divides $q+1$. The first case is the most subtle, because the Sylow 2-subgroup is non-abelian. The second case is the simplest and all $\mathcal{O}$-blocks of the group with non-trivial defect group are Morita equivalent to $\mathcal{O}$-blocks of N, or even to T for some of them (see Corollary 8.3.2). In the last case, all non-principal $\mathcal{O}$-blocks of G are Morita equivalent to blocks of N', or even of T' for some of them (see Propositions 8.1.4 and 8.2.2), while the principal $\mathcal{O}$-block is Rickard equivalent to the principal $\mathcal{O}$-block of N'. This can be seen by algebraic methods, valid for all abstract finite groups (see [Ric1], [Lin] and [Rou2]) but, in order to remain true to the spirit of this book and to illustrate Deligne-Lusztig theory, we construct these equivalences with the help of the cohomology $\mathbf{R}\Gamma_c(\mathbf{Y}, \mathcal{O})$ of the Drinfeld curve (with coefficients in $\mathcal{O}$). This construction was conjectured by Broué, Malle and Michel [BrMaMi, "Zyklotomische Heckealgebren", §1A] for an arbitrary reductive group, and shown in the case of the group $G = \mathrm{SL}_2(\mathbb{F}_q)$ by Rouquier [Rou1] (see Corollary 8.3.5). To be precise, Rouquier treats the case of the group $\mathrm{GL}_2(\mathbb{F}_q)$ but his method can be easily adapted to the group $\mathrm{SL}_2(\mathbb{F}_q)$.

Finally, in Chapter 10, we give a brief account of the representations in equal characteristic. The majority of this chapter is devoted to the construction of the simple modules. In accordance with the general theory of finite reductive groups, all simple modules are obtained by restriction from rational representations of the algebraic group $\mathbf{G} = \mathrm{SL}_2(\mathbb{F})$.

NOTATION – In the previous chapters, the irreducible characters of G have been parametrised, using a prime number ℓ different to p and an ℓ-adic field K. Once this parametrisation has been obtained, the irreducible characters may be viewed as $\mathbb{K}G$-modules, where $\mathbb{K}$ is any sufficiently large field of characteristic zero. We may also take $\mathbb{K}$ to be a finite extension of $\mathbb{Q}_p$.

In other words, if we do not make it explicit, we keep our field K, a sufficiently large extension of $\mathbb{Q}_\ell$, but *we allow the possibility that ℓ be equal to p.*

We denote by $\mathcal{O}$ the ring of integers of K over $\mathbb{Z}_\ell$ and $\mathfrak{l}$ denotes its maximal ideal. Set $k = \mathcal{O}/\mathfrak{l}$: it is a finite field of characteristic ℓ. If M is an $\mathcal{O}$-module, we denote by $\overline{M}$ its reduction modulo $\mathfrak{l}$, that is $\overline{M} = k \otimes_{\mathcal{O}} M$. If $m \in M$, the image of m in $\overline{M}$ will be denoted $\overline{m}$.

Chapter 6
More about Characters of G and of its Sylow Subgroups

The purpose of this chapter is to assemble some preliminary results (exclusively concerning irreducible characters) which will be useful in the study of modular representations. We describe central characters, congruences and the character tables of normalisers of Sylow subgroups. We make use of this information to verify the global McKay conjecture for G in all characteristics.

6.1. Central Characters

In order to determine the blocks of G, we will use Proposition B.1.5 applied to Table 6.1 below, which gives the central characters of G.

Table 6.1 shows that we will need some results about congruences modulo $\mathfrak{l}$ of cyclotomic numbers. To this end, denote by $\mu_{\ell'}(\mathcal{O})$ (respectively $\mu_{\ell^\infty}(\mathcal{O})$) the group of roots of unity of $\mathcal{O}$ which are of order prime to ℓ (respectively of order a power of ℓ) and denote by $\mu(\mathcal{O})$ the group of roots of unity of $\mathcal{O}$. Note that

$$\mu(\mathcal{O}) = \mu_{\ell'}(\mathcal{O}) \times \mu_{\ell^\infty}(\mathcal{O}).$$

It is well known that

(6.1.1) *the kernel of the canonical morphism* $\mu(\mathcal{O}) \to k^\times$ *is* $\mu_{\ell^\infty}(\mathcal{O})$.

We deduce the following result.

Lemma 6.1.2. *Let a and b be two roots of unity in $\mathcal{O}$ such that $a + a^{-1} \equiv b + b^{-1}$ $\bmod \mathfrak{l}$. Then $ab^{-1} \in \mu_{\ell^\infty}(\mathcal{O})$ or $ab \in \mu_{\ell^\infty}(\mathcal{O})$.*

Proof. The hypothesis implies that $\overline{a} + \overline{a}^{-1} = \overline{b} + \overline{b}^{-1}$. Now, $\overline{a}\,\overline{a}^{-1} = \overline{b}\,\overline{b}^{-1} = 1$, and therefore $\overline{a} \in \{\overline{b}, \overline{b}^{-1}\}$. The result then follows from 6.1.1. □

C. Bonnafé, *Representations of* $\mathrm{SL}_2(\mathbb{F}_q)$, Algebra and Applications 13,
DOI 10.1007/978-0-85729-157-8_6,

Table 6.1 Central characters of $G = \mathrm{SL}_2(\mathbb{F}_q)$

g	εI_2	$\mathbf{d}(a)$	$\mathbf{d}'(\xi)$	εu_τ
	$\varepsilon^2 = 1$	$a \in \mu_{q-1} \setminus \{\pm 1\}$	$\xi \in \mu_{q+1} \setminus \{\pm 1\}$	$\varepsilon \in \{\pm 1\}, \tau \in \{\pm\}$
$\lvert \mathrm{Cl}_G(g) \rvert$	1	q^2+q	q^2-q	$\dfrac{q^2-1}{2}$
$o(g)$	$o(\varepsilon)$	$o(a)$	$o(\xi)$	$p \cdot o(\varepsilon)$
$C_G(g)$	G	T	T'	$\{\pm I_2\} \times U$
1_G	1	$q(q+1)$	$q(q-1)$	$\dfrac{q^2-1}{2}$
St_G	1	$q+1$	$-(q-1)$	0
$\mathrm{R}(\alpha), \alpha^2 \neq 1$	$\alpha(\varepsilon)$	$q(\alpha(a)+\alpha(a)^{-1})$	0	$\dfrac{q-1}{2}\alpha(\varepsilon)$
$\mathrm{R}'(\theta), \theta^2 \neq 1$	$\theta(\varepsilon)$	0	$-q(\theta(\xi)+\theta(\xi)^{-1})$	$-\dfrac{q+1}{2}\theta(\varepsilon)$
$\mathrm{R}_\sigma(\alpha_0), \sigma = \pm$	$\alpha_0(\varepsilon)$	$2q\alpha_0(a)$	0	$(q-1)\alpha_0(\varepsilon)\dfrac{1+\sigma\tau\sqrt{q_0}}{2}$
$\mathrm{R}'_\sigma(\theta_0), \sigma = \pm$	$\theta_0(\varepsilon)$	0	$-2q\theta_0(\xi)$	$(q+1)\theta_0(\varepsilon)\dfrac{-1+\sigma\tau\sqrt{q_0}}{2}$

Corollary 6.1.3. *Let α and β be two linear characters of a finite abelian group Γ such that $\alpha(\gamma)+\alpha(\gamma)^{-1} \equiv \beta(\gamma)+\beta(\gamma)^{-1} \mod \mathfrak{l}$ for all $\gamma \in \Gamma$. Then $\alpha^{-1}\beta$ or $\alpha\beta$ is of order a power of ℓ.*

6.2. Global McKay Conjecture

Here we verify the global McKay conjecture (stated in Appendix B) in order to lay the groundwork for the study of more structural questions, like equivalences of categories. To establish equivalences between certain blocks of G and blocks of normalisers of Sylow subgroups, we will use Theorems B.2.5 and B.2.6. It is therefore necessary to understand the characters of such normalisers. Recall that N (respectively N', respectively B) is the normaliser of a Sylow ℓ-subgroup when ℓ is odd and divides $q-1$ (respectively ℓ is odd and divides $q+1$, respectively $\ell = p$).

6.2.1. *Characters of N*

We will always identify the groups T and μ_{q-1} using the isomorphism $\mathbf{d}$. If $\alpha \in T^\wedge$, we set

$$\chi_\alpha = \operatorname{Ind}_T^N \alpha.$$

As ${}^s\alpha = \alpha^{-1}$, a linear character α of T is N-invariant if and only if $\alpha^2 = 1$. Moreover, $\chi_\alpha = \chi_{\alpha^{-1}}$. Denote by 1_N the trivial character of N, ε the linear character of order 2 which is trivial on T, $\chi_{\alpha_0}^{\pm}$ the unique extension of α_0 to N such that $\chi_{\alpha_0}^{\pm}(s) = \pm\sqrt{\alpha_0(-1)}$ (recall that $s^2 = \mathbf{d}(-1)$). Then Clifford theory [Isa, Theorem 6.16] shows that

$$\operatorname{Irr} N = \{1_N, \varepsilon, \chi_{\alpha_0}^{+}, \chi_{\alpha_0}^{-}\} \,\dot{\cup}\, \{\chi_\alpha \mid \alpha \in [T^\wedge/\equiv], \alpha^2 \neq 1\}.$$

The character table of N is then easy to calculate and is given in Table 6.2 (we set $s_+ = s$ and $s_- = s\mathbf{d}(z_0)$, where z_0 is a non-square in $\mathbb{F}_q^\times$).

Table 6.2 Character table of N

g	εI_2 $\varepsilon \in \{\pm 1\}$	s_τ $\tau \in \{\pm\}$	$\mathbf{d}(a)$ $a \in \mu_{q-1} \setminus \{\pm 1\}$
$\lvert \mathrm{Cl}_N(g) \rvert$	1	$\frac{q-1}{2}$	2
$o(g)$	$o(\varepsilon)$	4	$o(a)$
$C_N(g)$	N	$\langle s_\tau \rangle$	T
1_N	1	1	1
ε	1	-1	1
$\chi_{\alpha_0}^{\sigma}, \sigma \in \{\pm\}$	$\alpha_0(\varepsilon)$	$\sigma\tau\sqrt{\alpha_0(-1)}$	$\alpha_0(a)$
$\chi_\alpha, \; \alpha^2 \neq 1$	$2\alpha(\varepsilon)$	0	$\alpha(a) + \alpha(a)^{-1}$

In particular,

(6.2.1) $$\lvert \operatorname{Irr} N \rvert = \frac{q+5}{2}.$$

6.2.2. *Characters of* N'

The character table of N' is obtained in essentially the same way as for that of N. If θ is a linear character of T', we set

$$\chi'_\theta = \mathrm{Ind}_{T'}^{N'} \theta.$$

As ${}^{s'}\theta = \theta^{-1}$, we conclude that $\chi'_\theta = \chi'_{\theta^{-1}}$ and that θ is N'-invariant if and only if $\theta^2 = 1$. Denote by $1_{N'}$ the trivial character of N', ε' the unique linear character of order 2 which is trivial on T' and $\chi'^{\pm}_{\theta_0}$ the unique extension of θ_0 with value $\pm\sqrt{\theta_0(-1)}$ on s'. Clifford theory [Isa, Theorem 6.16] shows that

$$\mathrm{Irr}\, N' = \{1_{N'}, \varepsilon', \chi'^{+}_{\theta_0}, \chi'^{-}_{\theta_0}\} \mathbin{\dot\cup} \{\chi'_\theta \mid \theta \in [(T')^\wedge/\equiv], \theta^2 \neq 1\}.$$

The character table of N' is then easily calculated and is given in Table 6.3 (we set $s'_+ = s'$ and $s'_- = s\mathbf{d}'(\xi_0)$, where ξ_0 is a non-square in μ_{q+1}).

Table 6.3 Character table of N'

g	εI_2	s'_τ	$\mathbf{d}'(\xi)$
	$\varepsilon \in \{\pm 1\}$	$\tau \in \{\pm\}$	$\xi \in \mu_{q+1} \setminus \{\pm 1\}$
$\lvert \mathrm{Cl}_{N'}(g) \rvert$	1	$\frac{q+1}{2}$	2
$o(g)$	$o(\varepsilon)$	4	$o(\xi)$
$C_{N'}(g)$	N'	$\langle s'_\tau \rangle$	T'
$1_{N'}$	1	1	1
ε'	1	-1	1
$\chi'^{\sigma}_{\theta_0}, \sigma \in \{\pm\}$	$\theta_0(\varepsilon)$	$\sigma\tau\sqrt{\theta_0(-1)}$	$\theta_0(\xi)$
$\chi'_\theta,\ \theta^2 \neq 1$	$2\theta(\varepsilon)$	0	$\theta(\xi) + \theta(\xi)^{-1}$

In particular,

(6.2.2) $$\lvert \mathrm{Irr}\, N' \rvert = \frac{q+7}{2}.$$

6.2.3. *Characters of B*

Denote by 1_+ (respectively 1_-) the trivial (respectively non-trivial) linear character of Z. Recall that χ_+ is a non-trivial linear character of $\mathbb{F}_q^+$ and that ψ_+ is the corresponding non-trivial linear character of U under the isomorphism $\mathbf{u}\colon \mathbb{F}_q^+ \xrightarrow{\sim} U$. We denote by ψ_- an irreducible component of the character Υ_- defined in 5.2. If $\sigma, \tau \in \{+,-\}$, we set

$$\chi_{\sigma,\tau}^B = \operatorname{Ind}_{ZU}^B 1_\sigma \boxtimes \psi_\tau.$$

One may easily verify that

$$\operatorname{Irr} B = \{\tilde{\alpha} \mid \alpha \in T^\wedge\} \dot{\cup} \{\chi_{+,+}^B, \chi_{+,-}^B, \chi_{-,+}^B, \chi_{-,-}^B\}$$

and that the character table of B is given by Table 6.4 (recall that $q_0 = \alpha_0(-1)q$).

Table 6.4 Character table of B

g	εI_2 $\varepsilon \in \{\pm 1\}$	$\mathbf{d}(a)$ $a \in \mathbb{F}_q^\times \setminus \{\pm 1\}$	$\varepsilon u_{\tau'}$ $\varepsilon, \tau' \in \{+,-\}$
$\lvert \mathrm{Cl}_B(g) \rvert$	1	q	$\dfrac{q-1}{2}$
$o(g)$	$o(\sigma)$	$o(a)$	$o(\sigma)\cdot p$
$C_B(g)$	B	T	ZU
$\tilde{\alpha},\quad \alpha \in T^\wedge$	$\alpha(\varepsilon)$	$\alpha(a)$	$\alpha(\varepsilon)$
$\chi_{\sigma,\tau'}^B, \sigma, \tau \in \{\pm\}$	$\dfrac{(q-1)1_\sigma(\varepsilon)}{2}$	0	$1_\sigma(\varepsilon)\dfrac{-1+\tau\tau'\sqrt{q_0}}{2}$

In particular,

(6.2.3) $$\lvert \operatorname{Irr} B \rvert = q+3.$$

6.2.4. *Normalisers of Sylow 2-Subgroups*

As the Sylow 2-subgroups of G are non-abelian, we will be content to state the following result, leaving its proof and the calculation of the character table as an exercise (see Exercise 6.1).

(6.2.4) *If S is a Sylow 2-subgroup of G, then* $|\operatorname{Irr}_{2'} \mathrm{N}_G(S)| = 4$.

6.2.5. *Verification of the Global McKay Conjecture*

We can now verify the following result.

Proposition 6.2.5. *If S is a Sylow ℓ-subgroup of G, then*

$$|\operatorname{Irr}_{\ell'} G| = |\operatorname{Irr}_{\ell'} \mathrm{N}_G(S)|.$$

Proof. We may assume that ℓ divides the order of G. Set

$$\mathscr{E} = \{\mathsf{R}(\alpha) \mid \alpha \in T^\wedge, \alpha^2 \neq 1\} \quad \text{and} \quad \mathscr{E}' = \{\mathsf{R}'(\theta) \mid \theta \in \mu_{q+1}^\wedge, \theta^2 \neq 1\}.$$

Note that

$$|\mathscr{E}| = \frac{q-3}{2} \quad \text{and} \quad |\mathscr{E}'| = \frac{q-1}{2}.$$

We have

$$\operatorname{Irr}_{\ell'}(G) = \begin{cases} \{1_G, \mathrm{St}_G, \mathsf{R}_+(\alpha_0), \mathsf{R}_-(\alpha_0)\} & \text{if } \ell = 2 \text{ and } q \equiv 1 \mod 4, \\ \{1_G, \mathrm{St}_G, \mathsf{R}'_+(\theta_0), \mathsf{R}'_-(\theta_0)\} & \text{if } \ell = 2 \text{ and } q \equiv 3 \mod 4, \\ \{1_G, \mathrm{St}_G, \mathsf{R}_+(\alpha_0), \mathsf{R}_-(\alpha_0)\} \,\dot{\cup}\, \mathscr{E} & \text{if } \ell \text{ is odd and divides } q-1, \\ \{1_G, \mathrm{St}_G, \mathsf{R}'_+(\theta_0), \mathsf{R}'_-(\theta_0)\} \,\dot{\cup}\, \mathscr{E}' & \text{if } \ell \text{ is odd and divides } q+1, \\ (\operatorname{Irr} G) \setminus \{\mathrm{St}_G\} & \text{if } \ell = p. \end{cases}$$

It follows that

$$|\operatorname{Irr}_{\ell'} G| = \begin{cases} 4 & \text{if } \ell = 2, \\ \dfrac{q+5}{2} & \text{if } \ell \text{ is odd and divides } q-1, \\ \dfrac{q+7}{2} & \text{if } \ell \text{ is odd and divides } q+1, \\ q+3 & \text{if } \ell = p. \end{cases}$$

Moreover, when S is abelian, it follows from a theorem of Ito [Isa, Theorem 6.14] that $|\mathrm{Irr}_{\ell'} \mathrm{N}_G(S)| = |\mathrm{Irr}\, \mathrm{N}_G(S)|$ because ℓ does not divide $|\mathrm{N}_G(S)/S|$. We have therefore verified the global McKay conjecture thanks to 6.2.1, 6.2.2, 6.2.3 and 6.2.4. □

Exercises

6.1. Let S be a Sylow 2-subgroup of G. Calculate the character table of $\mathrm{N}_G(S)$ and verify 6.2.4.

6.2. Show that, if ℓ divides $|G|$ and is different from p, then St_G is in the principal ℓ-block of G.

REMARK – On the other hand, the Steinberg character St_G is not in the principal p-block of G (in fact, it is alone in its p-block by B.2.4). □

6.3. Verify McKay's conjecture for $\widetilde{G} = \mathrm{GL}_2(\mathbb{F}_q)$.

6.4*. Determine, as a function of ℓ, the radical filtration of the kG-module $\mathrm{Ind}_B^G k$.

Chapter 7
Unequal Characteristic: Generalities

Hypotheses. *In this and the following two chapters, we assume that ℓ is a prime number different from p.*

The purpose of this chapter is to assemble results valid in all unequal characteristics. We determine, for example, the decomposition into $\mathcal{O}$-blocks as well as the Brauer correspondents. We also introduce modular (that is over $\mathcal{O}$, or even over $\mathbb{Z}_\ell$) and structural versions of Harish-Chandra and Deligne-Lusztig induction. These are functors between categories, rather than being simply maps between Grothendieck groups. The preliminary work on these two functors will be useful in the next chapter, where we study the equivalences of categories which they induce (which turn out to be either Morita or derived equivalences).

7.1. Blocks, Brauer Correspondents

7.1.1. Partition in ℓ-Blocks

Using Table 6.1 (central characters) together with Corollary 6.1.3, we can determine the partition of $\operatorname{Irr} G$ into ℓ-blocks. We will need the following notation: we denote by $T_{\ell'}$ and $T'_{\ell'}$ the maximal subgroups of T and T' of order prime to ℓ, so that

$$T = S_\ell \times T_{\ell'} \quad \text{and} \quad T' = S'_\ell \times T'_{\ell'}.$$

These isomorphisms allow us to identify $T^\wedge$ and $T'^\wedge$ with $S_\ell^\wedge \times T_{\ell'}^\wedge$ and $S_\ell'^\wedge \times T_{\ell'}'^\wedge$ respectively. On the other hand, we identify T' and μ_{q+1} via the isomorphism $\mathbf{d}'$. If α (respectively θ) is a linear character of $T_{\ell'}$ (respec-

C. Bonnafé, *Representations of* $\mathrm{SL}_2(\mathbb{F}_q)$, Algebra and Applications 13, DOI 10.1007/978-0-85729-157-8_7, © Springer-Verlag London Limited 2011

tively $T'_{\ell'}$), denote by $\mathscr{B}_\alpha$ (respectively $\mathscr{B}'_\theta$) the set formed by the irreducible components of all $\mathsf{R}(\alpha\lambda)$ (respectively $\mathsf{R}'(\theta\lambda)$), where $\lambda \in S_\ell^\wedge$ (respectively $\lambda \in S_\ell'^\wedge$). The following theorem describes the decomposition of $\operatorname{Irr} G$ into ℓ-blocks (recall that $\ell \neq p$).

Theorem 7.1.1. *If $\ell \neq p$, then:*

(a) *If $\alpha \in T_{\ell'}^\wedge$ is such that $\alpha^2 \neq 1$, then $\mathscr{B}_\alpha$ is an ℓ-block of G with defect group S_ℓ.*
(b) *If $\theta \in T_{\ell'}'^\wedge$ is such that $\theta^2 \neq 1$, then $\mathscr{B}'_\theta$ is an ℓ-block of G with defect group S'_ℓ.*
(c) *If ℓ is odd and divides $q-1$, then:*

(c1) *$\{\mathsf{R}'_+(\theta_0)\}$ and $\{\mathsf{R}'_-(\theta_0)\}$ are ℓ-blocks of G with defect group $1 = S'_\ell$ (note that $\mathscr{B}'_{\theta_0} = \{\mathsf{R}'_+(\theta_0), \mathsf{R}'_-(\theta_0)\}$).*
(c2) *$\mathscr{B}_1$ and $\mathscr{B}_{\alpha_0}$ are ℓ-blocks of G with defect group S_ℓ ($\mathscr{B}_1$ being the principal ℓ-block).*

(d) *If ℓ is odd and divides $q+1$, then:*

(d1) *$\{\mathsf{R}_+(\alpha_0)\}$ and $\{\mathsf{R}_-(\alpha_0)\}$ are ℓ-blocks of G with defect group $1 = S_\ell$ (note that $\mathscr{B}_{\alpha_0} = \{\mathsf{R}_+(\alpha_0), \mathsf{R}_-(\alpha_0)\}$).*
(d2) *$\mathscr{B}'_1$ and $\mathscr{B}'_{\theta_0}$ are ℓ-blocks of G with defect group S'_ℓ ($\mathscr{B}'_1$ being the principal ℓ-block).*

(e) *If $\ell = 2$, then $\mathscr{B}_1 \cup \mathscr{B}'_1$ is the principal ℓ-block of G, with defect group a Sylow 2-subgroup of G (note that $\mathscr{B}_1 \cap \mathscr{B}'_1 = \{1_G, \mathrm{St}_G\}$).*

Proof. This follows from a careful inspection of Table 6.1, Corollary 6.1.3 and the fact that q is invertible in $\mathscr{O}$. □

7.1.2. Brauer Correspondents

Let α (respectively θ) be a linear character of $T_{\ell'}$ (respectively $T'_{\ell'}$) such that $\alpha^2 \neq 1$ (respectively $\theta^2 \neq 1$) and denote by A_α (respectively A'_θ) the $\mathscr{O}$-block of G such that $\operatorname{Irr} KA_\alpha = \mathscr{B}_\alpha$ (respectively $\operatorname{Irr} KA'_\theta = \mathscr{B}'_\theta$). The principal $\mathscr{O}$-block of G will be denoted A_1 or A'_1. If moreover ℓ is *odd*, we denote by A_{α_0} (respectively A'_{θ_0}) the sum of the $\mathscr{O}$-blocks of G satisfying $\operatorname{Irr} KA_{\alpha_0} = \mathscr{B}_{\alpha_0}$ (respectively $\operatorname{Irr} KA'_{\theta_0} = \mathscr{B}'_{\theta_0}$).

REMARK – If ℓ is odd and divides $q-1$, then A_{α_0} is an $\mathscr{O}$-block with defect group S_ℓ while A'_{θ_0} is the sum of two $\mathscr{O}$-blocks with defect group $1 = S'_\ell$.

If ℓ is odd and divides $q+1$, then A_{α_0} is the sum of two $\mathscr{O}$-blocks with defect group $1 = S_\ell$ while A'_{θ_0} is an $\mathscr{O}$-block with defect group S'_ℓ.

If A is the sum of multiple $\mathscr{O}$-blocks of G with the same defect group S, we still use *Brauer correspondent* to refer to the sum of the associated Brauer correspondents (i.e. a sum of $\mathscr{O}$-blocks of $\mathsf{N}_G(S)$). □

If α (respectively θ) is a linear character of $T_{\ell'}$ (respectively $T'_{\ell'}$) such that $\alpha^2 \neq 1$ (respectively $\theta^2 \neq 1$), we denote by b_α (respectively b'_θ) the primitive central idempotent of $\mathcal{O}T$ (respectively $\mathcal{O}T'$) equal to

$$b_\alpha = \frac{1}{|T_{\ell'}|} \sum_{t \in T_{\ell'}} \alpha(t)\, t^{-1}$$

(respectively $$b'_\theta = \frac{1}{|T'_{\ell'}|} \sum_{t' \in T'_{\ell'}} \theta(t')\, t'^{-1} \qquad).$$

It is now straightforward to calculate the Brauer correspondents (the reader is referred to Exercise 1.6 for the calculation of the normalisers of subgroups of T and T'). These are the object of the following theorem, where the correspondents are given only for those $\mathcal{O}$-blocks with non-central defect group.

Theorem 7.1.2. *Let α and θ be linear characters of $T_{\ell'}$ and $T'_{\ell'}$ respectively. Let S be a Sylow 2-subgroup of G.*

(a) *If ℓ is odd and divides $q-1$, then:*

(a1) *If $\alpha^2 = 1$, then $\mathcal{O}Nb_\alpha$ is the Brauer correspondent of A_α.*
(a2) *If $\alpha^2 \neq 1$, then $\mathcal{O}N(b_\alpha + b_{\alpha^{-1}})$ is the Brauer correspondent of A_α.*

(b) *If ℓ is odd and divides $q+1$, then:*

(b1) *If $\theta^2 = 1$, then $\mathcal{O}N' b'_\theta$ is the Brauer correspondent of A'_θ.*
(b2) *If $\theta^2 \neq 1$, then $\mathcal{O}N'(b'_\theta + b'_{\theta^{-1}})$ is the Brauer correspondent of A'_θ.*

(c) *If $\ell = 2$, then the principal $\mathcal{O}$-block of $\mathrm{N}_G(S)$ is the Brauer correspondent of $A_1 = A'_1$. Moreover:*

(c1) *If $q \equiv 1 \mod 4$ and $\alpha \neq 1$, then $\mathcal{O}N(b_\alpha + b_{\alpha^{-1}})$ is the Brauer correspondent of A_α.*
(c2) *If $q \equiv 3 \mod 4$ and $\theta \neq 1$, then $\mathcal{O}N'(b'_\theta + b'_{\theta^{-1}})$ is the Brauer correspondent of A'_θ.*

REMARK – In statements (c1) and (c2) above, the condition $\alpha \neq 1$ (respectively $\theta \neq 1$) is equivalent to $\alpha^2 \neq 1$ (respectively $\theta^2 \neq 1$): indeed, in (c), we have $\ell = 2$, therefore α and θ are of odd order. □

Proof. We will prove (a), the other cases are treated in a similar manner and are left to the reader. Therefore suppose that ℓ is odd and divides $q-1$. Denote by i_α the primitive central idempotent in $\mathcal{O}G$ such that $B_\alpha = \mathcal{O}Gi_\alpha$. We will show that

$$\mathrm{Br}_{S_\ell}(\bar{i}_\alpha) = \begin{cases} \bar{b}_\alpha & \text{if } \alpha^2 = 1, \\ \bar{b}_\alpha + \bar{b}_{\alpha^{-1}} & \text{if } \alpha^2 \neq 1. \end{cases}$$

As $C_G(S_\ell) = T$, it is enough to calculate the coefficients of i_α on elements of the group T. If $t \in T$, the coefficient of t in i_α is equal to

$$\begin{cases} \frac{1}{|G|}\Big(1\times 1 + q\times 1 + (q+1)\times \sum\limits_{\substack{\lambda\in[S_\ell^\wedge/\equiv]\\ \lambda\neq 1}} \lambda(t)+\lambda(t^{-1})\Big) & \text{if } \alpha = 1, \\ \frac{1}{|G|}\Big(2\times \frac{q+1}{2}\times \alpha_0(t) + (q+1)\times \sum\limits_{\substack{\lambda\in[S_\ell^\wedge/\equiv]\\ \lambda\neq 1}} \alpha_0(t)(\lambda(t)+\lambda(t^{-1}))\Big) & \text{if } \alpha = \alpha_0, \\ \frac{q+1}{|G|}\sum\limits_{\lambda\in S_\ell^\wedge} \lambda(t)+\lambda(t)^{-1} & \text{if } \alpha^2 \neq 1. \end{cases}$$

This coefficient is therefore equal to

$$\begin{cases} \frac{(q+1)|S_\ell|}{|G|}\alpha(t)\delta_{t\in T_{\ell'}} & \text{if } \alpha^2 = 1, \\ \frac{(q+1)|S_\ell|}{|G|}(\alpha(t)+\alpha(t^{-1}))\delta_{t\in T_{\ell'}} & \text{if } \alpha^2 \neq 1, \end{cases}$$

which shows the desired result, because $(q+1)|S_\ell|/|G| = 1/q|T_{\ell'}| \equiv 1/|T_{\ell'}| \mod \mathfrak{l}$ (since $q \equiv 1 \mod \mathfrak{l}$). □

7.1.3. *Terminology*

If α (respectively θ) is a linear character of $T_{\ell'}$ (respectively $T'_{\ell'}$) such that $\alpha^2 \neq 1$ (respectively $\theta^2 \neq 1$), we will say that the block A_α (respectively A'_θ) is *nilpotent*. (The reader may verify that they are indeed nilpotent in the sense of Broué and Puig [BrPu, definition 1.1].)

If ℓ is odd, the (sum of) blocks A_{α_0} and A'_{θ_0} will be called *quasi-isolated*. (One may verify that this agrees with the definition of a quasi-isolated semi-simple element given in [Bon, §1.B] and the Jordan decomposition of blocks of reductive groups in unequal characteristic [BrMi, Theorem 2.2].)

REMARK – The notion of nilpotent block is algebraic (as it is defined for any block of any finite group) whereas the notion of quasi-isolated block is of more geometric nature (as it is defined only for finite reductive groups and requires Deligne-Lusztig theory). □

7.2. Modular Harish-Chandra Induction

The functors of Harish-Chandra induction and restriction are defined by

$$\begin{array}{rcl} \mathscr{R}\colon\ \mathbb{Z}_\ell T\text{-mod} & \longrightarrow & \mathbb{Z}_\ell G\text{-mod} \\ M & \longmapsto & \mathbb{Z}_\ell[G/U]\otimes_{\mathbb{Z}_\ell T} M \end{array}$$

and

$$\begin{array}{rcl} {}^*\mathscr{R}\colon\ \mathbb{Z}_\ell G\text{-mod} & \longrightarrow & \mathbb{Z}_\ell T\text{-mod} \\ M & \longmapsto & \mathbb{Z}_\ell[U\backslash G]\otimes_{\mathbb{Z}_\ell G} M. \end{array}$$

If Λ is a commutative $\mathbb{Z}_\ell$-algebra, we denote by $\mathscr{R}_\Lambda$ and ${}^*\mathscr{R}_\Lambda$ the extension of scalars to Λ of the functors $\mathscr{R}$ and ${}^*\mathscr{R}$ respectively. For example, if $\Lambda = K$, we obtain the functors $\mathscr{R}_K$ and ${}^*\mathscr{R}_K$ defined in Section 3.2.

Proposition 7.2.1. *The $(\mathbb{Z}_\ell G, \mathbb{Z}_\ell T)$-bimodule $\mathbb{Z}_\ell[G/U]$ is projective both as a left $\mathbb{Z}_\ell G$-module and as a right $\mathbb{Z}_\ell T$-module.*

Proof. Let $e_U = (1/q)\sum_{u\in U} u$. Then e_U is idempotent and belongs to $\mathbb{Z}_\ell U \subseteq \mathbb{Z}_\ell G$ as q is invertible in $\mathbb{Z}_\ell$. Now, $\mathbb{Z}_\ell[G/U] \simeq \mathbb{Z}_\ell G e_U$ as a $(\mathbb{Z}_\ell G, \mathbb{Z}_\ell T)$-bimodule. The proposition follows, as $\mathbb{Z}_\ell G$ is free both as a $\mathbb{Z}_\ell G$ and $\mathbb{Z}_\ell T$-module. □

Corollary 7.2.2. *The functors $\mathscr{R}$ and ${}^*\mathscr{R}$ are adjoint.*

We now turn to a study of the endomorphism algebra of the $\mathbb{Z}_\ell G$-module $\mathbb{Z}_\ell[G/U]$. To this end we revisit a part of Exercise 3.3, this time working over the ring $\mathbb{Z}_\ell$ rather than the field K. Firstly, note that the right action of T on G/U induces a morphism of $\mathbb{Z}_\ell$-algebras

$$\begin{array}{rcl} \mathbb{Z}_\ell T & \longrightarrow & \mathrm{End}_{\mathbb{Z}_\ell G}\,\mathbb{Z}_\ell[G/U] \\ t\in T & \longmapsto & (x\mapsto x\cdot t^{-1}). \end{array}$$

We will view the algebra $\mathrm{End}_{\mathbb{Z}_\ell G}\,\mathbb{Z}_\ell[G/U]$ as a $\mathbb{Z}_\ell T$-module via this morphism.

On the other hand, denote by $\mathscr{F}$ the unique $\mathbb{Z}_\ell$-linear endomorphism of $\mathbb{Z}_\ell[G/U]$ such that

$$\mathscr{F}(gU) = \sum_{u\in U} gusU$$

for all $g \in G$. Recall that $s = \begin{pmatrix} 0 & -1 \\ 1 & 0 \end{pmatrix} \in N \subseteq \mathrm{N}_G(T)$. It is easy to verify that, if $g \in G$, $t \in T$ and $x \in \mathbb{Z}_\ell[G/U]$, then

$$\mathscr{F}(g\cdot x\cdot t) = g\cdot\mathscr{F}(x)\cdot {}^s t. \tag{7.2.3}$$

In particular, $\mathscr{F} \in \mathrm{End}_{\mathbb{Z}_\ell G}\,\mathbb{Z}_\ell[G/U]$. Thanks to the endomorphisms induced by T and $\mathscr{F}$ we may describe the endomorphism algebra $\mathrm{End}_{\mathbb{Z}_\ell G}\,\mathbb{Z}_\ell[G/U]$ as a kind of "quadratic extension" of the group algebra $\mathbb{Z}_\ell T$.

Theorem 7.2.4. *Set* $E = \sum_{t \in T} t \in \mathbb{Z}_\ell T$. *Then:*

(a) *We have the following equality between elements of* $\mathrm{End}_{\mathbb{Z}_\ell G}\, \mathbb{Z}_\ell[G/U]$:

$$\mathscr{F}^2 = q\mathbf{d}(-1) + \mathscr{F} \cdot E.$$

Moreover, $\mathscr{F} \circ t = {}^s t \circ \mathscr{F}$ *for all* $t \in \mathbb{Z}_\ell T$.

(b) *If* $\tau, \tau' \in \mathbb{Z}_\ell T$ *satisfies* $\tau + \tau' \mathscr{F} = 0$, *then* $\tau = \tau' = 0$.

(c) $\mathrm{End}_{\mathbb{Z}_\ell G}\, \mathbb{Z}_\ell[G/U] = \mathbb{Z}_\ell T \oplus \mathbb{Z}_\ell T \cdot \mathscr{F}$.

(d) *If* Λ *is a commutative* $\mathbb{Z}_\ell$*-algebra, then* $\mathrm{End}_{\Lambda G}\, \Lambda[G/U] = \Lambda \otimes_{\mathbb{Z}_\ell} \mathrm{End}_{\mathbb{Z}_\ell G}\, \mathbb{Z}_\ell[G/U]$.

Proof. (a) Let $g \in G$. Then

$$\mathscr{F}^2(gU) = \sum_{u,v \in U} gusvsU.$$

Now, if $a, b \in \mathbb{F}_q$, then

$$g \begin{pmatrix} 1 & a \\ 0 & 1 \end{pmatrix} s \begin{pmatrix} 1 & b \\ 0 & 1 \end{pmatrix} sU = \begin{cases} g\mathbf{d}(-1)U & \text{if } b = 0, \\ g \begin{pmatrix} 1 & -a \\ 0 & 1 \end{pmatrix} s\mathbf{d}(b^{-1})U & \text{if } b \neq 0. \end{cases}$$

It follows that,

$$\mathscr{F}^2(gU) = qg\mathbf{d}(-1)U + \sum_{u \in U} gus\Big(\sum_{t \in T} t\Big)U.$$

Hence the result.

We now turn to (b), (c) and (d). By using the isomorphism of $(\Lambda G, \Lambda T)$-bimodules $\Lambda[G/U] \simeq \Lambda G e_U$. The map

$$\begin{array}{ccc} \mathrm{End}_{\Lambda G}\, \Lambda G e_U & \longrightarrow & (e_U \Lambda G e_U)^{\mathrm{opp}} \\ f & \longmapsto & f(e_U) \end{array}$$

is an isomorphism of Λ-algebras. Now the set $(e_U n e_U)_{n \in N}$ gives a Λ-basis of $e_U \Lambda G e_U$ by 1.1.7. Via this isomorphism, $\mathscr{F}$ corresponds to $q e_U s e_U$. Now, if $t \in T$ and $n \in N$, we have $(e_U n e_U)(e_U t e_U) = e_U n t e_U$, which shows (b), (c) and (d). □

In Chapter 8, we will use Theorem 7.2.4 to show that, if ℓ is odd and divides $q-1$, then $\mathrm{End}_{\mathbb{Z}_\ell G}\, \mathbb{Z}_\ell[G/U] \simeq \mathbb{Z}_\ell N$ (see Theorem 8.3.1): this arithmetic condition is related to Broué's conjecture.

7.3. Deligne-Lusztig Induction*

As G and μ_{q+1} act freely on the Drinfeld curve $\mathbf{Y}$ (see Propositions 2.1.2 and 2.1.3), it follows from Theorem A.1.5 that

(7.3.1) *the complex* $\mathsf{R\Gamma}_c(\mathbf{Y}, \mathbb{Z}_\ell)$ *is left and right perfect,*

that is, quasi-isomorphic to a bounded complex of $(\mathbb{Z}_\ell G, \mathbb{Z}_\ell \mu_{q+1})$-bimodules projective as left and right modules. In order to maintain consistent notation we will identify μ_{q+1} and T' via the isomorphism $\mathbf{d}'$ and view $\mathbf{Y}$ as a variety equipped with an action of $G \times (T' \rtimes \langle F \rangle_{\mathrm{mon}})$, with F acting on T' as conjugation by $s' \in N'$. In this way $\mathsf{R\Gamma}_c(\mathbf{Y}, \mathbb{Z}_\ell)$ will be viewed as a complex of $(\mathbb{Z}_\ell G, \mathbb{Z}_\ell T')$-bimodules, inducing a functor between the bounded derived categories

$$\begin{aligned} \mathscr{R}' \colon \mathsf{D}^b(\mathbb{Z}_\ell T') &\longrightarrow \mathsf{D}^b(\mathbb{Z}_\ell G) \\ C &\longmapsto \mathsf{R\Gamma}_c(\mathbf{Y}, \mathbb{Z}_\ell) \otimes_{\mathbb{Z}_\ell T'} C. \end{aligned}$$

If Λ is a commutative $\mathbb{Z}_\ell$-algebra we denote by $\mathscr{R}'_\Lambda$ the extension of scalars to Λ of the functor $\mathscr{R}'$. On the other hand, as $\mathbf{Y}$ is a smooth curve, it follows from A.1.4 that

(7.3.2) $H^i_c(\mathbf{Y}, \mathbb{Z}_\ell)$ *is a free* $\mathbb{Z}_\ell$*-module.*

Recall also that, by Theorem A.2.1, we have

(7.3.3) $H^2_c(\mathbf{Y}, \mathbb{Z}_\ell) = \mathbb{Z}_\ell$ (equipped with the trivial action of G)

and, by A.2.3, 7.3.2 and Theorem 2.2.2,

(7.3.4) $H^1_c(\mathbf{Y}, \mathbb{Z}_\ell)^G = 0.$

From these properties we will deduce the following result.

Proposition 7.3.5 (Rouquier). *There exist two projective* $\mathbb{Z}_\ell G$*-modules* P *and* Q *together with a morphism* $d \colon P \to Q$ *satisfying the following properties.*

(a) *In the derived category* $\mathsf{D}^b(\mathbb{Z}_\ell G)$*, we have an isomorphism*

$$\mathsf{R\Gamma}_c(\mathbf{Y}, \mathbb{Z}_\ell) \simeq_{\mathsf{D}} (0 \longrightarrow P \xrightarrow{\ d\ } Q \longrightarrow 0),$$

the term P *being in degree* 1.

(b) *The* $\mathbb{Z}_\ell G$*-module* Q *is the projective cover of* $\mathbb{Z}_\ell$ *(regarded as a trivial* $\mathbb{Z}_\ell G$*-module).*

(c) *The* $\mathbb{Z}_\ell G$*-module* P *does not admit a subquotient isomorphic to* $\mathbb{Z}_\ell$ *(or, equivalently,* $(KP)^G = 0$*).*

Proof. By 7.3.1 and 7.3.2, there exists projective $\mathbb{Z}_\ell G$-modules P and Q together with a morphism of $\mathbb{Z}_\ell G$-modules $d \colon P \to Q$ such that $\mathsf{R\Gamma}_c(\mathbf{Y}, \mathbb{Z}_\ell)$ is quasi-isomorphic to the complex

$$\mathscr{C} = (0 \longrightarrow P \xrightarrow{d} Q \longrightarrow 0).$$

Here, the terms P and Q are situated in degrees 1 and 2. By 7.3.3, we have $Q/\operatorname{Im} d \simeq \mathbb{Z}_\ell$. As Q is projective, we can write $Q = P_1 \oplus Q'$, where P_1 is the projective cover of $\mathbb{Z}_\ell$ and Q' is contained in $\operatorname{Im} d$.

The projectivity of Q' implies that the canonical surjective morphism $d^{-1}(Q') \to Q'$ splits: we denote by Q'' the $\mathbb{Z}_\ell G$-submodule of P which is the image of this section. It is isomorphic to Q' via the restriction of d. On the other hand, the $\mathbb{Z}_\ell$-module P/Q'' is torsion-free: if $x \in P$ satisfies $\ell x \in Q''$, then $d(\ell x) \in Q'$ and therefore $d(x) \in Q'$ as Q/Q' is torsion-free. This shows that $x \in d^{-1}(Q')$ and in conclusion we remark that $d^{-1}(Q')/Q'' \simeq d^{-1}(Q') \cap \operatorname{Ker} d$ is torsion-free.

As P/Q'' is torsion-free, the exact sequence of $\mathbb{Z}_\ell G$-modules $0 \to Q'' \to P \to P/Q'' \to 0$ splits over $\mathbb{Z}_\ell$, and therefore splits as a sequence of $\mathbb{Z}_\ell G$-modules as Q'' is projective, therefore relatively $(G,1)$-projective and therefore relatively $(G,1)$-injective [CuRe, Theorem 19.2]. Denote by P' a complement of Q'' in P and denote by $d' : P' \to P_1$ the composition of the morphism $P' \to Q \to P_1$, the last arrow being the projection $Q = P_1 \oplus Q' \to P_1$. The complex

$$0 \longrightarrow P \xrightarrow{d} Q \longrightarrow 0$$

is therefore homotopic to a complex

$$0 \longrightarrow P' \xrightarrow{d'} P_1 \longrightarrow 0.$$

This shows that we may (and will) assume that

$$Q = P_1.$$

All that remains is to verify the third assertion. We have an exact sequence of $\mathbb{Z}_\ell G$-modules

$$0 \longrightarrow H^1_c(\mathbf{Y}, \mathbb{Z}_\ell) \longrightarrow P \xrightarrow{d} Q \longrightarrow H^2_c(\mathbf{Y}, \mathbb{Z}_\ell) \longrightarrow 0,$$

which, after tensoring with the flat $\mathbb{Z}_\ell$-algebra K, yields a short exact sequence

$$0 \longrightarrow H^1_c(\mathbf{Y}) \longrightarrow KP \xrightarrow{d_K} KQ \longrightarrow H^2_c(\mathbf{Y}) \longrightarrow 0,$$

where $d_K : KP \to KQ$ is the canonical extension of d. Because the algebra KG is semi-simple, the functor *"invariants under G"* is exact, which gives us a final exact sequence

$$0 \longrightarrow H^1_c(\mathbf{Y})^G = 0 \longrightarrow (KP)^G \longrightarrow (KQ)^G \longrightarrow H^2_c(\mathbf{Y})^G = K \longrightarrow 0.$$

Now, $(KQ)^G = K$ as Q is the projective cover of $\mathbb{Z}_\ell$, which shows property (c). □

Encouraged by Proposition 7.3.5, we will study the endomorphism algebra of $\mathbf{R\Gamma}_c(\mathbf{Y}, \mathbb{Z}_\ell)$, viewed as an object in the derived category $\mathsf{D}^b(\mathbb{Z}_\ell G)$. We will commence with the following important result, which will also prove useful in the construction of equivalences of derived categories in the following chapter.

Corollary 7.3.6 (Rouquier). *If $i \neq 0$, then*

$$\mathsf{Hom}_{\mathsf{D}^b(\mathbb{Z}_\ell G)}(\mathbf{R\Gamma}_c(\mathbf{Y}, \mathbb{Z}_\ell), \mathbf{R\Gamma}_c(\mathbf{Y}, \mathbb{Z}_\ell)[i]) = 0.$$

Moreover, the $\mathbb{Z}_\ell$-module $\mathsf{End}_{\mathsf{D}^b(\mathbb{Z}_\ell G)} \mathbf{R\Gamma}_c(\mathbf{Y}, \mathbb{Z}_\ell)$ is torsion-free.

Proof. We keep the notation $(P,\ Q,\ d)$ used in the statement of Proposition 7.3.5 and denote by $\mathscr{C}$ the complex

$$\mathscr{C} = (0 \longrightarrow P \xrightarrow{\ d\ } Q \longrightarrow 0).$$

As P and Q are projective $\mathbb{Z}_\ell G$-modules, we have

$$\mathsf{Hom}_{\mathsf{D}^b(\mathbb{Z}_\ell G)}(\mathbf{R\Gamma}_c(\mathbf{Y}, \mathbb{Z}_\ell), \mathbf{R\Gamma}_c(\mathbf{Y}, \mathbb{Z}_\ell)[i]) = \mathsf{Hom}_{\mathsf{K}^b(\mathbb{Z}_\ell G)}(\mathscr{C}, \mathscr{C}[i]).$$

We begin with the first assertion. In virtue of the previous equality, it suffices to show that

$$\mathsf{Hom}_{\mathsf{K}^b(\mathbb{Z}_\ell G)}(\mathscr{C}, \mathscr{C}[i]) = 0$$

for all $i \neq 0$. As the result is evident for $|i| \geqslant 2$, we assume that $|i| = 1$.

• Suppose that $i = 1$. Then $\mathsf{Hom}_{\mathsf{C}^b(\mathbb{Z}_\ell G)}(\mathscr{C}, \mathscr{C}[1]) \simeq \mathsf{Hom}_{\mathbb{Z}_\ell G}(P, Q)$. Therefore, let $\varphi \in \mathsf{Hom}_{\mathbb{Z}_\ell G}(P, Q)$ and view φ as a morphism of complexes

$$\begin{array}{lccccccc}
\mathscr{C} & & 0 & \longrightarrow & P & \xrightarrow{\ d\ } & Q & \longrightarrow 0 \\
 & & & \downarrow & & \downarrow{\scriptstyle\varphi} & & \downarrow \\
\mathscr{C}[1] & 0 & \longrightarrow & P & \xrightarrow{\ d\ } & Q & \longrightarrow & 0.
\end{array}$$

Suppose first that φ is not contained in the image of d. Then φ induces a surjection $KP \to H^2_c(\mathbf{Y}) \simeq K$. It follows that $(KP)^G \neq 0$, which is impossible by Proposition 7.3.5(c).

Therefore $\mathsf{Im}\,\varphi \subseteq \mathsf{Im}\,d$. The projectivity of P then allows one to construct a morphism $\varphi'\colon P \to P$ such that $\varphi = d \circ \varphi'$. This shows that the morphism of complexes φ is homotopic to zero.

• Now suppose that $i = -1$. To give a morphism of complexes $\varphi\colon \mathscr{C} \to \mathscr{C}[-1]$ is the same as giving a morphism of $\mathbb{Z}_\ell G$-modules $\varphi\colon Q \to P$ such that $\mathsf{Im}\,\varphi \subseteq \mathsf{Ker}\,d$ and $\mathsf{Im}\,d \subseteq \mathsf{Ker}\,\varphi$:

$$\begin{array}{lccccccc}
\mathscr{C} & 0 & \longrightarrow & P & \xrightarrow{d} & Q & \longrightarrow & 0 \\
 & \downarrow & & & \swarrow{\scriptstyle\varphi} & & & \downarrow \\
\mathscr{C}[-1] & 0 & \longrightarrow & P & \xrightarrow{d} & Q & \longrightarrow & 0.
\end{array}$$

Therefore φ factorises to give a morphism $Q/\operatorname{Im} d \to P$ and hence to a morphism $\mathbb{Z}_\ell \to P$. This last morphism is necessarily zero by virtue of Proposition 7.3.5(c). Therefore $\varphi = 0$.

We now turn to the last assertion, concerning the absence of torsion in the endomorphism algebra $\mathbf{R\Gamma}_c(\mathbf{Y},\mathbb{Z}_\ell)$. So let $\varphi \in \operatorname{End}_{C^b(\mathbb{Z}_\ell G)}(\mathscr{C})$ be such that $\ell\varphi$ is homotopic to zero. We have to show that φ itself is homotopic to zero. Let $\alpha\colon P \to P$ and $\beta\colon Q \to Q$ be the morphisms of $\mathbb{Z}_\ell G$-modules such that φ is equal to the following morphism

$$\begin{array}{lccccccc}
\mathscr{C} & 0 & \longrightarrow & P & \xrightarrow{d} & Q & \longrightarrow & 0 \\
 & & & \downarrow{\scriptstyle\alpha} & & \downarrow{\scriptstyle\beta} & & \\
\mathscr{C} & 0 & \longrightarrow & P & \xrightarrow{d} & Q & \longrightarrow & 0.
\end{array}$$

By the hypotheses, there exists a morphism of $\mathbb{Z}_\ell G$-modules $\rho\colon Q \to P$ such that $\ell\alpha = \rho \circ d$ and $\ell\beta = d \circ \rho$. As P and Q are torsion-free it is enough to show that $\operatorname{Im}\rho \subseteq \ell P$.

Denote by $\bar\rho\colon Q \to P/\ell P$ the composition of $\rho\colon Q \to P$ with the canonical projection $P \to P/\ell P$. Then $\bar\rho$ is zero on $\operatorname{Im} d$, and therefore $\bar\rho$ factorises to give a morphism $\tilde\rho\colon H^2_c(\mathbf{Y},\mathbb{Z}_\ell) \to P/\ell P$. If $\tilde\rho$ is non-zero, then $P/\ell P$ contains an $\mathbb{F}_\ell G$-submodule isomorphic to $\mathbb{F}_\ell$. But $P/\ell P = \mathbb{F}_\ell P$ is a projective $\mathbb{F}_\ell G$-module. Therefore if it contains the trivial module in its socle, then it contains the projective cover of the trivial module as a direct summand. This implies that P admits the $\mathbb{Z}_\ell G$-module Q as a direct summand, and therefore $P^G \neq 0$, contradicting Proposition 7.3.5(c). Therefore $\tilde\rho = 0$ and $\bar\rho = 0$, which concludes the proof of the corollary. □

Denote by $H^\bullet_c(\mathbf{Y},\mathbb{Z}_\ell)$ the graded $\mathbb{Z}_\ell G$-module $\oplus_{i \geqslant 0} H^i_c(\mathbf{Y},\mathbb{Z}_\ell)$ (in fact, we have $H^\bullet_c(\mathbf{Y},\mathbb{Z}_\ell) = H^1_c(\mathbf{Y},\mathbb{Z}_\ell) \oplus H^2_c(\mathbf{Y},\mathbb{Z}_\ell)$). The previous proposition has the following consequence.

Corollary 7.3.7 (Rouquier). *The natural morphism of $\mathcal{O}$-algebras*

$$\operatorname{End}_{D^b(\mathbb{Z}_\ell G)} \mathbf{R\Gamma}_c(\mathbf{Y},\mathbb{Z}_\ell) \longrightarrow \operatorname{End}^{\mathrm{gr}}_{\mathbb{Z}_\ell G} H^\bullet_c(\mathbf{Y},\mathbb{Z}_\ell)$$

is injective (here, $\operatorname{End}^{\mathrm{gr}}_?(-)$ denotes the algebra of ***graded*** *endomorphisms).*

Proof. By Corollary 7.3.6, the $\mathbb{Z}_\ell$-algebra $\operatorname{End}_{D^b(\mathbb{Z}_\ell G)} \mathbf{R\Gamma}_c(\mathbf{Y},\mathbb{Z}_\ell)$ injects naturally into $\operatorname{End}_{D^b(\mathbb{Q}_\ell G)} \mathbf{R\Gamma}_c(\mathbf{Y},\mathbb{Q}_\ell)$. But as $\mathbb{Q}_\ell G$ is semi-simple, this last algebra is isomorphic to $\operatorname{End}^{\mathrm{gr}}_{\mathbb{Q}_\ell G} H^\bullet_c(\mathbf{Y},\mathbb{Q}_\ell)$. The commutativity of the diagram

$$\begin{array}{ccc} \mathrm{End}_{D^b(\mathbb{Z}_\ell G)} \mathbf{R\Gamma}_c(\mathbf{Y},\mathbb{Z}_\ell) & \longrightarrow & \mathrm{End}^{\mathrm{gr}}_{\mathbb{Z}_\ell G} H^\bullet_c(\mathbf{Y},\mathbb{Z}_\ell) \\ \downarrow & & \downarrow \\ \mathrm{End}_{D^b(\mathbb{Q}_\ell G)} \mathbf{R\Gamma}_c(\mathbf{Y},\mathbb{Q}_\ell) & \xrightarrow{\sim} & \mathrm{End}^{\mathrm{gr}}_{\mathbb{Q}_\ell G} H^\bullet_c(\mathbf{Y},\mathbb{Q}_\ell) \end{array}$$

implies the corollary. □

Having established these preliminary results, we now turn to the question of determining the endomorphism algebra $\mathrm{End}_{D^b(\mathbb{Z}_\ell G)} \mathbf{R\Gamma}_c(\mathbf{Y},\mathbb{Z}_\ell)$, using similar techniques to those used in the determination of $\mathrm{End}_{\mathbb{Z}_\ell G} \mathbb{Z}_\ell[G/U]$. The right action of T' on $\mathbf{Y}$ commutes with the left action of G, and we therefore have a canonical morphism of $\mathbb{Z}_\ell$-algebras $\mathbb{Z}_\ell T' \to \mathrm{End}_{D^b(\mathbb{Z}_\ell G)} \mathbf{R\Gamma}_c(\mathbf{Y},\mathbb{Z}_\ell)$. Using this morphism we view $\mathrm{End}_{D^b(\mathbb{Z}_\ell G)} \mathbf{R\Gamma}_c(\mathbf{Y},\mathbb{Z}_\ell)$ as a $\mathbb{Z}_\ell T'$-module. On the other hand, the Frobenius endomorphism F of the Drinfeld curve induces an endomorphism, denoted $\mathscr{F}'$, of the complex $\mathbf{R\Gamma}_c(\mathbf{Y},\mathbb{Z}_\ell)$. As the action of G commutes with that of F, this yields an element of $\mathrm{End}_{D^b(\mathbb{Z}_\ell G)} \mathbf{R\Gamma}_c(\mathbf{Y},\mathbb{Z}_\ell)$. It follows from the relations between the actions of F, G and T on $\mathbf{Y}$ that

$$\mathscr{F}' \circ (g,t') = (g,t'^{-1}) \circ \mathscr{F}' \tag{7.3.8}$$

for all $(g,t') \in G \times T'$. The following theorem describes the structure of the algebra $\mathrm{End}_{D^b(\mathbb{Z}_\ell G)} \mathbf{R\Gamma}_c(\mathbf{Y},\mathbb{Z}_\ell)$ in an analogous fashion to the description of $\mathrm{End}_{\mathbb{Z}_\ell G} \mathbb{Z}_\ell[G/U]$ in Theorem 7.2.4.

Theorem 7.3.9 (Rouquier). *Set $E' = \sum_{t' \in T'} t' \in \mathscr{O}T'$. Then:*

(a) *We have the following equality between elements of $\mathrm{End}_{D^b(\mathbb{Z}_\ell G)} \mathbf{R\Gamma}_c(\mathbf{Y},\mathbb{Z}_\ell)$:*

$$\mathscr{F}'^2 = -q\mathbf{d}'(-1) + \mathscr{F}' \circ E'.$$

Moreover, $\mathscr{F}' \circ t' = {}^{s'}t' \circ \mathscr{F}'$ for all $t' \in \mathbb{Z}_\ell T'$.

(b) *If τ, $\tau' \in \mathbb{Z}_\ell T'$ satisfy $\tau + \tau'\mathscr{F}' = 0$, then $\tau = \tau' = 0$.*

(c) *We have*

$$\mathrm{End}_{D^b(\mathbb{Z}_\ell G)} \mathbf{R\Gamma}_c(\mathbf{Y},\mathbb{Z}_\ell) = \mathbb{Z}_\ell T' \oplus \mathbb{Z}_\ell T' \cdot \mathscr{F}'.$$

(d) *If Λ is a commutative $\mathbb{Z}_\ell$-algebra, then*

$$\mathrm{End}_{D^b(\Lambda G)} \mathbf{R\Gamma}_c(\mathbf{Y},\Lambda) \simeq \Lambda \otimes_{\mathbb{Z}_\ell} \mathrm{End}_{D^b(\mathbb{Z}_\ell G)} \mathbf{R\Gamma}_c(\mathbf{Y},\mathbb{Z}_\ell).$$

Proof. (a) By virtue of Corollary 7.3.7, it is enough to verify this equality on the cohomology groups $H^1_c(\mathbf{Y},\mathbb{Z}_\ell)$ and $H^2_c(\mathbf{Y},\mathbb{Z}_\ell)$. But, by 7.3.2, it is enough to verify this equality on the cohomology groups $H^1_c(\mathbf{Y},K) = H^1_c(\mathbf{Y})$ and $H^2_c(\mathbf{Y},K) = H^2_c(\mathbf{Y})$. This is done using the calculation of the eigenvalues of F and F^2 completed in Section 4.4. We have:

- On $H_c^2(\mathbf{Y})$, μ_{q+1} acts trivially (therefore E' acts as multiplication by $q+1$) and $\mathscr{F}'$ acts as multiplication by q (see 4.4.1).
- On $H_c^1(\mathbf{Y})e_1$, μ_{q+1} acts trivially (therefore E' acts as multiplication by $q+1$) and $\mathscr{F}'$ acts as multiplication by 1 (see 4.4.3).
- On $H_c^1(\mathbf{Y})e_{\theta_0}$, an element $\xi \in \mu_{q+1}$ acts as multiplication by $\theta_0(\xi)$ (therefore E' acts as multiplication by 0) and $\mathscr{F}'^2$ acts as multiplication by $-q\theta_0(-1)$ (see 4.4.5).
- If θ is a linear character of μ_{q+1} such that $\theta^2 \neq 1$, then, on $H_c^1(\mathbf{Y})(e_\theta + e_{\theta^{-1}})$, an element $\xi \in \mu_{q+1}$ acts as multiplication by $\theta(\xi)$ (therefore E' acts as multiplication by 0) and $\mathscr{F}'^2$ acts as multiplication by $-q\theta(-1)$ (see 4.4.5).

The result follows immediately from these observations.

(b) If $\tau + \tau'\mathscr{F}' = 0$, then this equality is still true on $H_c^1(\mathbf{Y})$ and $H_c^2(\mathbf{Y})$. If $\theta \in T'^\wedge$, we denote by $\hat{\theta}$ its linear extension to the group algebra KT'.

On $H_c^2(\mathbf{Y})$, $\tau+\tau'\mathscr{F}'$ acts as multiplication by $\hat{1}(\tau)+q\hat{1}(\tau')$. On $H_c^1(\mathbf{Y})e_1$, $\tau+\tau'\mathscr{F}'$ acts as multiplication by $\hat{1}(\tau)-\hat{1}(\tau')$. As a consequence, $\hat{1}(\tau)=\hat{1}(\tau')=0$.

On the space $H_c^1(\mathbf{Y})e_{\theta_0}$, τ and τ' act as multiplication by $\hat{\theta}_0(\tau)$ and $\hat{\theta}_0(\tau')$ respectively, while $\mathscr{F}'$ has two eigenvalues, $\sqrt{-\theta_0(-1)q}$ and $-\sqrt{-\theta_0(-1)q}$. Hence, $\hat{\theta}_0(\tau)+\hat{\theta}_0(\tau)\sqrt{-\theta_0(-1)q} = \hat{\theta}_0(\tau)-\hat{\theta}_0(\tau)\sqrt{-\theta_0(-1)q} = 0$, which implies that $\hat{\theta}_0(\tau)=\hat{\theta}_0(\tau')=0$.

We now study the action of $\tau+\tau'\mathscr{F}'$ on $H_c^1(\mathbf{Y})(e_\theta + e_{\theta^{-1}})$ if $\theta^2 \neq 1$. Let $(v_1,\dots,v_r)$ be a basis of $H_c^1(\mathbf{Y})e_\theta$. Then $(\mathscr{F}'(v_1),\dots,\mathscr{F}'(v_r))$ is a basis of $H_c^1(\mathbf{Y})e_{\theta^{-1}}$. In the basis $(v_1,\dots,v_r,\mathscr{F}'(v_1),\dots,\mathscr{F}'(v_r))$ of $H_c^1(\mathbf{Y})(e_\theta+e_{\theta^{-1}})$, the matrix of $\mathscr{F}'$ has the form

$$\mathscr{F}' \mapsto \begin{pmatrix} \mathbf{0} & -\theta(-1)q\mathrm{I}_r \\ \mathrm{I}_r & \mathbf{0} \end{pmatrix}$$

while an element τ'' of KT' has matrix

$$\tau'' \mapsto \mathrm{diag}(\underbrace{\hat{\theta}(\tau''),\dots,\hat{\theta}(\tau'')}_{r \text{ times}}, \underbrace{\hat{\theta^*}(\tau''),\dots,\hat{\theta^*}(\tau'')}_{r \text{ times}}).$$

The equality $\tau+\tau'\mathscr{F}'=0$ therefore implies $\hat{\theta}(\tau)=\hat{\theta}(\tau')=0$.

Hence we have shown that $\hat{\Theta}(\tau)=\hat{\Theta}(\tau')=0$ for all linear characters Θ of T'. Hence, $\tau=\tau'=0$.

(c) Set $\tilde{A} = \mathrm{End}_{D^b(\mathbb{Z}_\ell G)} \mathbf{R\Gamma}_c(\mathbf{Y},\mathbb{Z}_\ell)$ and $A = \mathbb{Z}_\ell T' \oplus \mathbb{Z}_\ell T' \cdot \mathscr{F}' \subseteq \tilde{A}$. We want to show that $A=\tilde{A}$. Note first that, by Proposition 7.3.6, the $\mathscr{O}$-modules A and $\tilde{A}$ are torsion-free. On the other hand, we claim that $KA = K\tilde{A}$. As K is flat over $\mathbb{Z}_\ell$, $K\tilde{A} = \mathrm{End}_{D^b(KG)} \mathbf{R\Gamma}_c(\mathbf{Y},K)$, so it follows from (b) that it is enough to show that

$$\dim_K \mathrm{End}_{D^b(KG)} \mathbf{R\Gamma}_c(\mathbf{Y},K) = 2\cdot|T'|.$$

But

$$\dim_K \operatorname{End}_{D^b(KG)} \mathbf{R\Gamma}_c(\mathbf{Y},K) = \langle [H^1_c(\mathbf{Y})]_{KG}, [H^1_c(\mathbf{Y})]_{KG}\rangle_G + \langle [H^2_c(\mathbf{Y})]_{KG}, [H^2_c(\mathbf{Y})]_{KG}\rangle_G.$$

As $\langle [H^1_c(\mathbf{Y})]_{KG}, [H^2_c(\mathbf{Y})]_{KG}\rangle_G = 0$, it is enough to show that

$$\langle \mathrm{R}'(\mathrm{reg}_{\mu_{q+1}}), \mathrm{R}'(\mathrm{reg}_{\mu_{q+1}})\rangle_{KG} = 2\cdot |T'|,$$

which follows immediately from the Mackey formula 4.2.1. We have therefore shown the claim, namely that $KA = K\tilde{A}$.

On the other hand, if we set, for $\varphi \in K\tilde{A}$,

$$\mathscr{T}(\varphi) = \frac{1}{|G|}(\mathrm{Tr}(\varphi, H^2_c(\mathbf{Y})) - \mathrm{Tr}(\varphi, H^1_c(\mathbf{Y}))) \quad \in K.$$

The key to the proof of (c) is the following lemma.

Lemma 7.3.10. *If $\varphi \in \tilde{A}$, then $\mathscr{T}(\varphi) \in \mathbb{Z}_\ell$. Moreover, the restriction of $\mathscr{T}$ to a morphism $\mathscr{T}: A \to \mathbb{Z}_\ell$ is a symmetrising form.*

Proof (of Lemma 7.3.10). As the complex $\mathbf{R\Gamma}_c(\mathbf{Y},\mathbb{Z}_\ell)$ is perfect as an object of $\mathrm{D}^b(\mathbb{Z}_\ell G)$ (see 7.3.1), it suffices to show that, if P is a projective $\mathbb{Z}_\ell G$-module and if $\varphi \in \mathrm{End}_{\mathbb{Z}_\ell G}(P)$, then $\mathrm{Tr}(\varphi, KP)/|G| \in \mathscr{O}$. To show this we may assume that P is indecomposable. Then there exists an idempotent $e \in \mathscr{O}G$ such that $P \simeq \mathbb{Z}_\ell Ge$ and we may assume that $P = \mathbb{Z}_\ell Ge$. If we denote by $\kappa\colon \mathbb{Z}_\ell G \to \mathbb{Z}_\ell$ the symmetrising form such that $\kappa(g) = \delta_{g=1}$ for all $g \in G$, then it is easy to verify that $\mathrm{Tr}(\varphi, KP) = |G|\kappa(\varphi(e))$. This shows the first claim.

We turn to the second claim. Denote by $\mathscr{B}$ the $\mathscr{O}$-basis of A obtained by the concatenation of $(t')_{t'\in T'}$ and $(t'\cdot\mathscr{F}')_{t'\in T'}$ and set $M = (\mathscr{T}(bb'))_{b,b'\in\mathscr{B}}$. It is enough to show that $\det M \in \mathbb{Z}_\ell^\times$. Now, for all t'_1, $t'_2 \in T'$, we have

$$\mathscr{T}(t'_1t'_2) = \begin{cases} 0 & \text{if } t'_1t'_2 \neq 1, \\ \dfrac{1}{q} & \text{if } t'_1t'_2 = 1, \end{cases}$$

$$\mathscr{T}(t'_1t'_2\mathscr{F}') = |\mathbf{Y}^{t'_1t'_2F}| = 0$$

and

$$\mathscr{T}(t'_1\mathscr{F}'t'_2\mathscr{F}') = |\mathbf{Y}^{t'_1t'^{-1}_2F^2}| = \begin{cases} 0 & \text{if } t'_1t'^{-1}_2 \neq -\mathrm{I}_2, \\ 1 & \text{if } t'_1t'^{-1}_2 = -\mathrm{I}_2. \end{cases}$$

Indeed, the first equality follows from Table 5.2, while the second and third follow from the Lefschetz fixed-point theorem combined with, respectively, 2.3.1 and Theorem 2.3.2. Hence, $\det(M) = \pm(1/q)^{|T'|} \in \mathscr{O}^\times$, as required. □

We now show why (c) follows from Lemma 7.3.10. Let $(a_1,\dots,a_d)$ be a $\mathbb{Z}_\ell$-basis of A and let $(a^*_1,\dots,a^*_d)$ be the dual $\mathbb{Z}_\ell$-basis of A for the form $\mathscr{T}$. Let

$a \in \tilde{A}$. Then, by $(*)$, there exists $\lambda_1, \ldots, \lambda_d \in K$ such that $a = \lambda_1 a_1 + \cdots + \lambda_d a_d$. By duality we obtain $\lambda_i = \mathscr{T}(aa_i^*)$. Now $aa_i^* \in \tilde{A}$ as $\tilde{A}$ is an algebra and so $\lambda_i = \mathscr{T}(aa_i^*) \in \mathscr{O}$ by Lemma 7.3.10, therefore $a \in A$. Hence $A = \tilde{A}$.

(d) follows immediately from the fact that $\mathbf{R\Gamma}_c(\mathbf{Y}, \mathbb{Z}_\ell)$ is perfect. □

In Chapter 8, we will use the previous theorem to show that, if ℓ is odd and divides $q+1$, then $\mathrm{End}_{\mathbb{Z}_\ell G}\, \mathbf{R\Gamma}_c(\mathbf{Y}, \mathbb{Z}_\ell) \simeq \mathbb{Z}_\ell N'$ (see Theorem 8.3.4).

Exercises

7.1. Complete the proof of 7.1.2.

7.2. We identify the $\mathbb{Z}_\ell G$-modules $\mathbb{Z}_\ell[G/B]$ with the space of T-invariants (for the right action) in $\mathbb{Z}_\ell[G/U]$. Show that $\mathscr{F}$ restricts to an endomorphism of $\mathbb{Z}_\ell[G/B]$. We will denote by $\mathscr{F}_B$ this restriction.

Denote by I the identity endomorphism of $\mathbb{Z}_\ell[G/B]$. Show that $\mathscr{F}_B^2 = q\mathrm{I} + (q-1)\mathscr{F}_B$ and that

$$\mathrm{End}_{\mathbb{Z}_\ell G}\, \mathbb{Z}_\ell[G/B] = \mathbb{Z}_\ell \cdot \mathrm{I} \oplus \mathbb{Z}_\ell \cdot \mathscr{F}_B.$$

7.3 (Howlett-Lehrer, Dipper-Du). Denote by U^- the subgroup of G formed by *lower* triangular matrices. Let R be a ring in which p is invertible. Identify the (RG, RT)-bimodules $R[G/U]$ and $R[G/U^-])$ with RGe_U and RGe_{U^-} respectively. Show that the map

$$\begin{array}{ccc} RGe_{U^-} & \longrightarrow & RGe_U \\ a & \longmapsto & ae_U \end{array}$$

is an isomorphism of (RG, RT)-bimodules. (***Hint:*** As both bimodules are free over R of the same rank, it is sufficient to show surjectivity, which follows from the formula

$$qe_U e_{U^-} e_U = e_U + \sum_{u \in U^- \setminus \{\mathrm{I}_2\}} e_U u e_U$$

and the fact that $U^- \setminus \{\mathrm{I}_2\} \subseteq TUsU$.)

7.4. Let R be a commutative ring in which p is invertible. If $\alpha\colon T \to R^\times$ is a linear character, we denote by R_α the RT-module defined as follows: the underlying R-module is R itself and an element $t \in T$ acts as multiplication by $\alpha(t)$. Use the previous exercise (or the automorphism $\mathscr{F}$, extended to the ring R) to show that $R[G/U] \otimes_{RT} R_\alpha \simeq R[G/U] \otimes_{RT} R_{\alpha^{-1}}$.

7.5. Show that, if V is a $\mathbb{Z}_\ell T$-module, then ${}^*\mathscr{R}(\mathscr{R}(V)) \simeq V \oplus {}^sV$, where ${}^sV$ denotes the $\mathbb{Z}_\ell T$-module with underlying $\mathbb{Z}_\ell$-module V but on which an element $t \in T$ acts as ${}^st = t^{-1}$.

Chapter 8
Unequal Characteristic: Equivalences of Categories

Hypothesis. *In this chapter, as in the next and the previous chapters, we assume that ℓ is a prime number different from p.*

The purpose of this chapter is to verify Broué's abelian defect conjecture (see Subsection B.2.2). In the case of non-principal blocks (which all have abelian defect groups), the equivalences of categories predicted by Broué's conjecture are always Morita equivalences (see Sections 8.1 and 8.2). While it is possible to obtain this result using Brauer trees and Brauer's theorem B.4.2, we give instead a concrete construction of these equivalences using Harish-Chandra and Deligne-Lusztig induction. In the case of principal blocks, treated in Section 8.3, the situation is more interesting. If ℓ is odd and divides $q-1$, then the principal block is Morita equivalent to its Brauer correspondent and Harish-Chandra induction induces an equivalence. If ℓ is odd and divides $q+1$, then the principal block is Rickard equivalent to its Brauer correspondent and Deligne-Lusztig induction induces the required equivalence. If $\ell = 2$, the situation is more complicated: when $q \equiv \pm 3 \mod 8$, then the principal block of G is Rickard equivalent to its Brauer correspondent; when $q \equiv \pm 1 \mod 8$, the derived category of the principal block is equivalent to the derived category of an A_∞-algebra. These two final results are due to Gonard [Go].

The last section is dedicated to *Alvis-Curtis duality*, viewed as an endofunctor of the homotopy category $\mathsf{K}^b(\mathbb{Z}_\ell G)$ or of the derived category $\mathsf{D}^b(\mathbb{Z}_\ell G)$. For an arbitrary finite reductive group it was shown by Cabanes and Rickard [CaRi] that this duality is an equivalence of the derived category. More recently, Okuyama [Oku3] improved this result by showing that it was in fact an equivalence of the homotopy category (see also the work of Cabanes [Ca] for a simplified treatment of Okuyama's theorem). We give a very concrete proof of Okuyama's result in the case of our small group $G = \mathrm{SL}_2(\mathbb{F}_q)$.

C. Bonnafé, *Representations of* $\mathrm{SL}_2(\mathbb{F}_q)$, Algebra and Applications 13,
DOI 10.1007/978-0-85729-157-8_8, © Springer-Verlag London Limited 2011

8.1. Nilpotent Blocks

8.1.1. Harish-Chandra Induction

Let α be a linear character of T such that $\alpha^2 \neq 1$. We will show the following.

Proposition 8.1.1. *Harish-Chandra induction $\mathcal{R}_{\mathcal{O}}$ induces a Morita equivalence between A_α and $\mathcal{O}Tb_\alpha$.*

Proof. Denote by f_α the primitive central idempotent of $\mathcal{O}G$ such that $A_\alpha = \mathcal{O}Gf_\alpha$. Set $M = \mathcal{O}[G/U]b_\alpha$. Then M is an $(\mathcal{O}G, \mathcal{O}Tb_\alpha)$-bimodule. Moreover, the irreducible factors of the KG-module KM are the irreducible characters of KA_α. Moreover, $f_\alpha M = M$ and therefore M is in fact an $(A_\alpha, \mathcal{O}Tb_\alpha)$-bimodule, which is projective as a left and right module by Proposition 7.2.1. We would like to show that the functor $M \otimes_{\mathcal{O}Tb_\alpha} -$ is an equivalence of categories.

By virtue of Broué's theorem B.2.5, it is enough to show that the functor $KM \otimes_{KTb_\alpha} -$ induces a bijection between $\operatorname{Irr} KTb_\alpha$ and $\operatorname{Irr} KA_\alpha$. As $\operatorname{Irr}(KTb_\alpha) = \{\alpha\lambda \mid \lambda \in S_\ell^\wedge\}$, this amounts to showing that $\operatorname{Irr} KA_\alpha = \{\mathrm{R}(\alpha\lambda) \mid \lambda \in S_\ell^\wedge\}$ which is precisely the definition of A_α. □

Corollary 8.1.2. *The $\mathcal{O}$-algebras A_α and $\mathcal{O}N(b_\alpha + b_{\alpha^{-1}})$ are Morita equivalent.*

Proof. By Proposition 8.1.1, it is enough to show that the $\mathcal{O}$-algebras $A = \mathcal{O}N(b_\alpha + b_{\alpha^{-1}})$ and $A' = \mathcal{O}Tb_\alpha$ are Morita equivalent. Set $P = \mathcal{O}Nb_\alpha$. Then P is an (A, A')-bimodule which is projective as a left and right module. We would like to show that the functor $P \otimes_{A'} -$ is an equivalence of categories.

By virtue of Broué's theorem B.2.5, it is enough to show that the functor $KP \otimes_{KA'} -$ induces a bijection between $\operatorname{Irr} KA'$ and $\operatorname{Irr} KA$. However this follows from the (easily verified) fact that $\operatorname{Irr} KA = \{\chi_{\alpha\lambda} \mid \lambda \in S_\ell^\wedge\}$ and $[KP \otimes_{KA'} K_{\alpha\lambda}]_{KN} = \chi_{\alpha\lambda}$ for all $\lambda \in S_\ell^\wedge$. □

COMMENTARY – When S_ℓ is not contained in Z (which occurs when ℓ is odd and divides $q-1$ or when $\ell = 2$ and $q \equiv 1 \mod 4$), then $\mathrm{N}_G(S_\ell) = N$ by Exercise 1.6 and $\mathcal{O}N(b_\alpha + b_{\alpha^{-1}})$ is the Brauer correspondent of A_α (by Theorem 7.1.2). As a consequence, Corollary 8.1.2 shows that Broué's conjecture (see Appendix B) is verified in a stronger form. Indeed, the predicted equivalence of derived categories is induced by a Morita equivalence.

Another consequence of Theorem 8.1.1 is that

$$A_\alpha \simeq \mathrm{Mat}_{q+1}(\mathcal{O}Tb_\alpha) \simeq \mathrm{Mat}_{q+1}(\mathcal{O}S_\ell), \tag{8.1.3}$$

which agrees with structure theorems for nilpotent blocks (for the case of an abelian defect group, see for example [BrPu, §1]). □

8.1.2. Deligne-Lusztig Induction*

Let θ be a linear character of T' such that $\theta^2 \neq 1$. We will show the following result.

Proposition 8.1.4. *Deligne-Lusztig induction $\mathscr{R}'_\theta$ induces a Morita equivalence between A'_θ and $\mathscr{O}Tb'_\theta$.*

Proof. Denote by f'_θ the primitive central idempotent of $\mathscr{O}G$ such that $A'_\theta = \mathscr{O}Gf'_\theta$. Set $M' = H^1_c(\mathbf{Y}, \mathscr{O})b'_\theta$. Then M' is an $(\mathscr{O}G, \mathscr{O}T'b'_\theta)$-bimodule and the irreducible factors of the KG-module KM' are the irreducible characters of KA'_θ. Furthermore, $f'_\theta M' = M'$ and therefore M' is in fact an $(A'_\theta, \mathscr{O}T'b'_\theta)$-bimodule. We would like to show that the functor $M' \otimes_{\mathscr{O}T'b'_\theta} -$ is an equivalence of categories.

We first show that M' is projective as a left and right module. By 7.3.1, the complex $f'_\theta \mathbf{R\Gamma}_c(\mathbf{Y}, \mathscr{O})b'_\theta$ is perfect as a left and right complex. Its i-th cohomology group is $f'_\theta H^i_c(\mathbf{Y}, \mathscr{O})b'_\theta$, which is non-zero only when $i = 1$. Therefore $f'_\theta \mathbf{R\Gamma}_c(\mathbf{Y}, \mathscr{O})b'_\theta$ is quasi-isomorphic to a complex of bimodules consisting of one non-zero term $f'_\theta H^1_c(\mathbf{Y}, \mathscr{O})b'_\theta = M'$ occurring in degree 1. Therefore M' is projective as a left and right module.

By virtue of Broué's theorem B.2.5, it is enough to show that the functor $KM' \otimes_{KT'b'_\theta} -$ gives a bijection between $\operatorname{Irr} KT'b'_\theta$ and $\operatorname{Irr} KA'_\theta$. As $\operatorname{Irr}(KT'b'_\theta) = \{\theta\lambda \mid \lambda \in S'^{\wedge}_\ell\}$, this amounts to saying that $\operatorname{Irr} KA'_\theta = \{\mathrm{R}'(\theta\lambda) \mid \lambda \in S'^{\wedge}_\ell\}$, which is precisely the definition of A'_θ. □

Corollary 8.1.5. *The $\mathscr{O}$-algebras A'_θ and $\mathscr{O}N'(b'_\theta + b'_{\theta^{-1}})$ are Morita equivalent.*

Proof. The proof follows the same lines as that of Corollary 8.1.2, the goal being to show that the $(\mathscr{O}N'(b'_\theta + b'_{\theta^{-1}}), \mathscr{O}T'b'_\theta)$-bimodule $\mathscr{O}N'b'_\theta$ induces a Morita equivalence. □

COMMENTARY – When S'_ℓ is not contained in Z (which occurs when ℓ is odd and divides $q+1$ or when $\ell = 2$ and $q \equiv 3 \mod 4$), then $\mathrm{N}_G(S'_\ell) = N'$ by Exercise 1.6 and $\mathscr{O}N'(b'_\theta + b'_{\theta^{-1}})$ is the Brauer correspondent of A'_θ (by Theorem 7.1.2). We conclude that Corollary 8.1.5 shows Broué's conjecture (see Appendix B) in a stronger form. Indeed, the predicted equivalence of derived categories is induced by a Morita equivalence.

Another consequence of Theorem 8.1.4 is that

$$A'_\theta \simeq \mathrm{Mat}_{q-1}(\mathscr{O}T'b_\theta) \simeq \mathrm{Mat}_{q-1}(\mathscr{O}S'_\ell), \tag{8.1.6}$$

which agrees with structure theorems for nilpotent blocks (for the case of an abelian defect group, see for example [BrPu, §1]). □

8.2. Quasi-Isolated Blocks

Hypothesis. *In this and only this section we assume that ℓ is odd (and different from p).*

This hypothesis is necessary for the existence of quasi-isolated blocks.

8.2.1. *Harish-Chandra Induction*

In this subsection we will show the following result.

Proposition 8.2.1. *Harish-Chandra induction $\mathcal{R}_{\mathcal{O}}$ induces a Morita equivalence between the $\mathcal{O}$-algebras A_{α_0} and $\mathcal{O}Nb_{\alpha_0}$.*

Proof. Denote by $\sqrt{q}$ a root of q in $\mathcal{O}$ (recall that $\mathcal{O}$ is sufficiently large). Set $M = \mathcal{O}[G/U]b_{\alpha_0}$. As in the proof of Proposition 8.1.1, M is an $(A_{\alpha_0}, \mathcal{O}Tb_{\alpha_0})$-bimodule. Moreover, by 7.2.3, M is stable under the endomorphism $\mathcal{F}$. Denote by $\mathcal{S}$ the restriction of $\mathcal{F}/\sqrt{q}$ to M. As the element $E = \sum_{t \in T} t$ acts as multiplication by 0 on M, we have

$$\mathcal{S}^2 = \mathbf{d}(-1)$$

by Theorem 7.2.4(a). Therefore, by 7.2.3, we can extend the right $\mathcal{O}T$-module structure on M to an $\mathcal{O}N$-module structure, letting s act as the automorphism $\mathcal{S}$. We may then view M as an $(A_{\alpha_0}, \mathcal{O}Nb_{\alpha_0})$-bimodule.

As N/T is of order 2 and ℓ is odd, it follows from Proposition 7.2.1 that M is projective as a left and right module. By virtue of Broué's Theorem B.2.5, it is enough to show that $\mathrm{End}_{A_{\alpha_0}} M \simeq \mathcal{O}Nb_{\alpha_0}$ (via the canonical morphism). This fact is a consequence of Theorem 7.2.4(c). □

COMMENTARY – If ℓ is odd and divides $q-1$, then $\mathcal{O}Nb_{\alpha_0}$ is the Brauer correspondent of A_{α_0} (see Theorem 7.1.2), and Proposition 8.2.1 shows that Broué's conjecture (see Appendix C) is true in a stronger form: the predicted equivalence of derived categories is induces by a Morita equivalence. □

8.2.2. *Deligne-Lusztig Induction**

In this subsection we will show the following result.

Proposition 8.2.2. *Deligne-Lusztig induction $\mathcal{R}'$ induces a Morita equivalence between the $\mathbb{Z}_\ell$-algebras A'_{θ_0} and $\mathcal{O}N'b'_{\theta_0}$.*

Proof. The proof is similar to that of Proposition 8.2.1. First, we define $M' = H^1_c(\mathbf{Y}, \mathcal{O})b'_{\theta_0}$. It is an $(A'_{\theta_0}, \mathcal{O}T'b'_{\theta_0})$-bimodule which is projective as a left and right module by the same argument as in the proof of Proposition 8.1.4. We may then give it the structure of an $(A'_{\theta_0}, \mathcal{O}N'b'_{\theta_0})$-bimodule, letting s' act as $\mathcal{F}'/\sqrt{-q}$ (by 7.3.8 and Theorem 7.3.9(a)). This bimodule is still projective as a left and right module, because N'/T' is of order 2 and ℓ is odd. It then results from Theorem 7.3.9(c) that $\mathrm{End}_{A'_{\theta_0}}(M') \simeq \mathcal{O}N'b'_{\theta_0}$. We may then conclude, thanks to Theorem B.2.5. □

COMMENTARY – If ℓ is odd and divides $q+1$, then $\mathcal{O}N'b'_{\theta_0}$ is the Brauer correspondent of A'_{θ_0} (see Theorem 7.1.2), and Proposition 8.2.2 shows Broué's conjecture (see Appendix B) in a stronger form. Indeed, the predicted equivalence of derived categories is induced by a Morita equivalence. □

8.3. The Principal Block

As described in the introduction to this chapter, the case of the principal block is the most interesting. Here we give a proof of the results described in the introduction when ℓ is odd. When $\ell = 2$, we will content ourselves with stating the results of Gonard [Go] without proof. In the richest case for which we give a complete treatment, that is when ℓ is odd and divides $q+1$, the equivalence predicted by Broué's conjecture will be constructed using the complex of cohomology $\mathbf{R\Gamma}_c(\mathbf{Y}, \mathcal{O})$. For this we will make crucial use of the results of the previous chapter, most notably the description of the algebra of endomorphisms of this complex.

8.3.1. *The Case when ℓ is Odd and Divides $q-1$*

We begin by showing a result of a "global" nature which refines Theorem 7.2.4.

Theorem 8.3.1. *If ℓ is odd and divides $q-1$, then* $\mathrm{End}_{\mathbb{Z}_\ell G}\,\mathbb{Z}_\ell[G/U] \simeq \mathbb{Z}_\ell N$.

Proof. As $q \equiv 1 \mod \ell$ and ℓ is odd, the polynomial $X^2 - q$ is split over $\mathbb{F}_\ell$. Therefore, by Hensel's lemma, there exists an element $\sqrt[+]{q}$ of $\mathbb{Z}_\ell$ such that $\sqrt[+]{q} \equiv 1 \mod \ell\mathbb{Z}_\ell$ and $(\sqrt[+]{q})^2 = q$. Set

$$\mathcal{S} = \frac{1}{\sqrt[+]{q}}(\mathcal{F} - \frac{1}{2}E)\cdot(1 - \frac{\sqrt[+]{q}-1}{(\sqrt[+]{q}+1)(q+1)}E),$$

and recall that $E = \sum_{t\in T} t \in \mathbb{Z}_\ell T$. As ℓ is odd and $\sqrt[+]{q} \equiv 1 \mod \ell\mathbb{Z}_\ell$, $\mathcal{S}$ belongs to $\mathrm{End}_{\mathbb{Z}_\ell G}\,\mathbb{Z}_\ell[G/U]$.

On the other hand a tedious but straightforward calculation shows that $\mathscr{S}^2 = \mathbf{d}(-1)$ and $\mathscr{S}t = t^{-1}\mathscr{S}$ for all $t \in T$. As a consequence, the right $\mathbb{Z}_\ell T$-module structure on $\mathbb{Z}_\ell[G/U]$ may be extended to a right $\mathbb{Z}_\ell N$-module structure by letting s act as $\mathscr{S}$. Moreover, $1 - \frac{\sqrt[+]{q}-1}{(\sqrt[+]{q}+1)(q+1)} E \in (\mathscr{O}T)^\times$ because, if $\lambda \in \ell\mathbb{Z}_\ell$, then the inverse of the element $1+\lambda E \in \mathscr{O}T$ is $1 - \frac{\lambda}{1+(q-1)\lambda} E \in \mathscr{O}T$. Theorem 7.2.4(c) then shows that $\mathrm{End}_{\mathbb{Z}_\ell G} \mathbb{Z}_\ell[G/U] = \mathbb{Z}_\ell T \oplus \mathbb{Z}_\ell T \cdot \mathscr{S} \simeq \mathbb{Z}_\ell N$, the isomorphism being induced by the right action of $\mathbb{Z}_\ell N$ on $\mathbb{Z}_\ell[G/U]$. □

Corollary 8.3.2. *If ℓ is odd and divides $q-1$, then Harish-Chandra induction induces a Morita equivalence between the $\mathscr{O}$-algebras $\oplus_{\alpha \in [T^\wedge/\equiv]} A_\alpha$ and $\mathscr{O}N$.*

Proof. Set

$$A = \bigoplus_{\alpha \in [T^\wedge/\equiv]} A_\alpha.$$

It follows from 3.2.13 that the irreducible factors of $K[G/U]$ are exactly the elements of $\mathrm{Irr}\, KA$. The result then follows from Theorem 8.3.1 and Broué's theorem B.2.5. □

COMMENTARY – When ℓ is odd and divides $q-1$ the Morita equivalence of the previous corollary induces an equivalence between A_α and its Brauer correspondent, after cutting by block idempotents. Thanks to Corollary 8.3.2, we rediscover when $\alpha \neq 1$ the results of Corollary 8.1.2 and Proposition 8.2.1 (which were valid in all unequal characteristic) by a more direct route. The reader may verify that the Morita equivalences constructed in these corollaries agree with those constructed in Corollary 8.3.2. □

Corollary 8.3.3. *If ℓ is odd and divides $q-1$, then Harish-Chandra induction induces a Morita equivalence between the principal $\mathscr{O}$-block of G and that of N.*

REMARK – Even when ℓ is odd and divides $q-1$, the $\mathscr{O}$-blocks A_{α_0} and $\mathscr{O}Tb_{\alpha_0}$ are not Morita equivalent. Similarly, the $\mathscr{O}$-blocks A_1 and $\mathscr{O}Tb_1$ are not Morita equivalent. There is even no equivalence of derived categories, because $|\mathrm{Irr}\, KA_?| = |S_\ell| + 1 \neq |S_\ell| = |\mathrm{Irr}\, KTb_?|$ when $? \in \{1, \alpha_0\}$ (see Remark B.2.7). □

8.3.2. *The Case when ℓ is Odd and Divides $q+1$**

We begin by showing a result of a "global" nature which refines Theorem 7.3.9.

Theorem 8.3.4. *If ℓ is odd and divides $q+1$, then* $\mathrm{End}_{D^b(\mathbb{Z}_\ell G)} \mathbf{R\Gamma}_c(\mathbf{Y}, \mathbb{Z}_\ell) \simeq \mathbb{Z}_\ell N'$.

Proof. The proof is analogous to that of Theorem 8.3.1. Indeed, set

$$\mathscr{S}' = \frac{1}{\sqrt[+]{-q}}(\mathscr{F}' - \frac{1}{2}E') \cdot (1 - \frac{\sqrt[+]{-q}-1}{(\sqrt[+]{-q}+1)(-q+1)} E'),$$

where we recall that $E' = \sum_{t' \in T'} t' \in \mathbb{Z}_\ell T'$ and $\sqrt[+]{-q}$ denotes a square root of $-q$ in $\mathbb{Z}_\ell$ such that $\sqrt[+]{-q} \equiv 1 \mod \ell\mathbb{Z}_\ell$ (this root exists because $-q \equiv 1 \mod \ell$).

One then shows that $\mathscr{S}'^2 = \mathbf{d}'(-1)$ and $\mathscr{S}' t' = t'^{-1} \mathscr{S}'$ for all $t' \in T'$. Thanks to Theorem 7.3.9(c) this allows us to show that

$$\mathrm{End}_{D^b(\mathbb{Z}_\ell G)} \mathbf{R\Gamma}_c(\mathbf{Y}, \mathbb{Z}_\ell) \simeq \mathbb{Z}_\ell N',$$

the isomorphism being obtained by letting $s' \in N'$ act through the automorphism $\mathscr{S}'$. □

Corollary 8.3.5 (Rouquier). *If ℓ is odd and divides $q+1$, then Deligne-Lusztig induction induces a Rickard equivalence between $\oplus_{\theta \in [T'^\wedge/\equiv]} A'_\theta$ and $\mathscr{O}N'$.*

Proof. The proof consists of two steps. In the first step we construct, using the building blocks at our disposal, a complex of $(\mathscr{O}G, \mathscr{O}N')$-bimodules. This is the most delicate step. The second step consists of showing that this complex verifies the conditions of Theorem B.2.6, which is almost a formality using the results at our disposal.

First step: construction of the complex. By Appendix A (section A.1), the complex $\mathbf{R\Gamma}_c(\mathbf{Y}, \mathscr{O})$ is a well-defined element in the category of complexes of bimodules $C^b(\mathscr{O}G, \mathscr{O}T' \rtimes \langle F \rangle_{\mathrm{mon}})$. Moreover, it is homotopic, in the category $K^b(\mathscr{O}G, \mathscr{O}T')$, to a bounded complex $\mathscr{C}$ of $(\mathscr{O}G, \mathscr{O}T')$-bimodules of finite type.

After eliminating direct factors homotopic to zero, we may suppose that

$\mathscr{C}$ is a complex of $(\mathscr{O}G, \mathscr{O}T')$-bimodules without a direct factor homotopic to zero.

This property has the following consequence.

Lemma 8.3.6. *In the algebra $\mathrm{End}_{C^b(\mathscr{O}G, \mathscr{O}T')} \mathscr{C}$ the two-sided ideal of morphisms homotopic to zero is contained in the radical of $\mathrm{End}_{C^b(\mathscr{O}G, \mathscr{O}T')} \mathscr{C}$.*

Proof (of Lemma 8.3.6). Denote by $\mathscr{A} = \mathrm{End}_{C^b(\mathscr{O}G, \mathscr{O}T')} \mathscr{C}$, $\mathscr{I}$ the two-sided ideal of morphisms which are homotopic to zero, and $\mathscr{R}$ the radical of $\mathscr{A}$. If $\mathscr{I}$ is not contained in $\mathscr{R}$, then $(\mathscr{I} + \mathscr{R})/\mathscr{R}$ is a *non-zero* two-sided ideal in the finite-dimensional semi-simple algebra $\mathscr{A}/\mathscr{R}$. In particular, $(\mathscr{I} + \mathscr{R})/\mathscr{R}$ contains a non-zero idempotent $\bar{e}$ of $\mathscr{A}/\mathscr{R}$. The theorem on lifting idempotents [The, Theorem 3.1 (h)] implies that there exists an idempotent $e \in \mathscr{I}$ whose image in $\mathscr{A}/\mathscr{R}$ is $\bar{e}$. It follows that $e\mathscr{C}$ is a direct factor of $\mathscr{C}$ and $e\colon e\mathscr{C} \to e\mathscr{C}$ is the identity, and therefore $e\mathscr{C}$ is homotopic to zero. This contradicts our hypothesis. Hence $\mathscr{I}$ is contained in $\mathscr{R}$ as claimed. □

Denote by $f\colon \mathscr{C} \to \mathbf{R\Gamma}_c(\mathbf{Y}, \mathscr{O})$ and $f'\colon \mathbf{R\Gamma}_c(\mathbf{Y}, \mathscr{O}) \to \mathscr{C}$ two morphisms of complexes which are mutually inverse in the category $K^b(\mathscr{O}G, \mathscr{O}T')$. Denote by $\tilde{S}'$ the element $\mathscr{O}T' \rtimes \langle F \rangle_{\mathrm{mon}}$ which is equal to

$$\tilde{S}' = \frac{1}{\sqrt[+]{-q}}(F' - \frac{1}{2}E') \cdot (1 - \frac{\sqrt[+]{-q}-1}{(\sqrt[+]{-q}+1)(-q+1)}E'),$$

by analogy with the proof of Theorem 8.3.4.

We then define $S' : \mathscr{C} \to \mathscr{C}$ by

$$S' = f' \circ \tilde{S}' \circ f,$$

where $\tilde{S}'$ is viewed as a morphism $\tilde{S}' : \mathbf{R}\Gamma_c(\mathbf{Y}, \mathscr{O}) \to \mathbf{R}\Gamma_c(\mathbf{Y}, \mathscr{O})$. Then, if $(g, t') \in G \times T'$, we have

$$(g, t') \circ S' = S' \circ (g, {}^{s'}t').$$

Moreover, as shown during the proof of Theorem 8.3.4, the element $S'^2 - \mathbf{d}'(-1)$ in $\mathrm{End}_{C^b(\mathscr{O}G, \mathscr{O}T')} \mathscr{C}$ is homotopic to zero. Set $h = \mathbf{d}'(-1)S'^2 - \mathrm{Id}_{\mathscr{C}}$. Then

$$S'^2 = \mathbf{d}'(-1)(1+h),$$

where $h \in \mathrm{End}_{C^b(\mathscr{O}G, \mathscr{O}T')} \mathscr{C}$ is homotopic to zero. By Lemma 8.3.6, h belongs to the radical of $\mathrm{End}_{C^b(\mathscr{O}G, \mathscr{O}T')} \mathscr{C}$ and we may therefore extract a square root of $1+h$ thanks to the standard formal series for $\sqrt{1+X}$. This series converges because $\ell \neq 2$. Therefore there exists $h' \in \mathrm{End}_{C^b(\mathscr{O}G, \mathscr{O}T')} \mathscr{C}$ which is homotopic to zero and such that $(1+h')^2 = 1+h$.

We now define $\sigma' : \mathscr{C} \to \mathscr{C}$ by $\sigma' = S' \circ (1+h')$. Then $\sigma'^2 = \mathbf{d}'(-1)$, which shows that we can equip $\mathscr{C}$ with the structure of a complex of $(\mathscr{O}G, \mathscr{O}N')$-bimodules, letting s' act by σ'. The equality $\sigma' \circ t' = {}^{s'}t' \circ \sigma'$ may be easily verified (for $t' \in T'$).

Second step: verification of the criteria of Theorem B.2.6. Let us define

$$A' = \bigoplus_{\theta \in [T'^\wedge/\equiv]} A'_\theta$$

and denote by e' the primitive central idempotent of $\mathscr{O}G$ such that $A' = \mathscr{O}Ge'$. Set

$$\mathscr{C}' = e'\mathscr{C}.$$

Then $\mathscr{C}'$ is a complex of $(A', \mathscr{O}N')$-bimodules which is quasi-isomorphic to $\mathscr{C}$ because e' acts as the identity on cohomology groups by 7.3.2. By Corollary 7.3.6, we have

$$\mathrm{Hom}_{D^b(A', \mathscr{O}N')}(\mathscr{C}', \mathscr{C}'[i]) = 0$$

for all $i \neq 0$. Moreover, every irreducible character of KA' is an irreducible factor of $H^\bullet(K\mathscr{C}') = H^1_c(\mathbf{Y}) \oplus H^2_c(\mathbf{Y})$ and the natural morphism

$$\mathscr{O}N' \to (\mathrm{End}_{D^b(\mathscr{O}G)} \mathscr{C}')^{\mathrm{opp}}$$

is an isomorphism by Theorem 8.3.4. The proof of the corollary is now complete by virtue of Theorem B.2.6. □

Corollary 8.3.7. *If ℓ is odd and divides $q+1$, then Deligne-Lusztig induction induces a Rickard equivalence between the principal $\mathcal{O}$-block of G and that of N'.*

REMARK – Even when ℓ is odd and divides $q+1$, the $\mathcal{O}$-blocks A'_{θ_0} and $\mathcal{O}T'b'_{\theta_0}$ are not Morita equivalent. Similarly, the $\mathcal{O}$-blocks A'_1 and $\mathcal{O}T'b'_1$ are not Morita equivalent. There is even no equivalence of derived categories, because $|\operatorname{Irr} KA'_?| = |S'_\ell| + 1 \neq |S'_\ell| = |\operatorname{Irr} KT'b'_?|$ where $? \in \{1, \theta_0\}$ (see Remark B.2.7). □

8.3.3. *The Case when $\ell = 2$**

In this subsection we will content ourselves to mention without proof two results of Gonard [Go]. Denote by S a Sylow 2-subgroup of G. Even though S is not abelian, the following result remains valid.

Theorem 8.3.8 (Gonard). *If $\ell = 2$ and $q \equiv \pm 3 \mod 8$ the principal $\mathcal{O}$-blocks of G and $\mathrm{N}_G(S)$ are Rickard equivalent.*

COMMENTARY – If $\ell = 2$ and $q \equiv 3 \mod 8$, recall that $|S| = 8$ and that $|\mathrm{N}_G(S)| = 24$ by Theorem 1.4.3. In particular, $S/Z \simeq \mathbb{Z}/2\mathbb{Z} \times \mathbb{Z}/2\mathbb{Z}$ is abelian and is a Sylow 2-subgroup of G/Z. As predicted by Broué's conjecture, it is possible to construct a Rickard equivalence between the principal blocks of G/Z and $\mathrm{N}_{G/Z}(S/Z) = \mathrm{N}_G(S)/Z$. Moreover, this equivalence may be lifted to give the Rickard equivalence of Theorem 8.3.8.

To be precise, Gonard works with the group $\widetilde{G} = \mathrm{GL}_2(\mathbb{F}_q)$, from which it is easy to deduce Theorem 8.3.8. □

Set

$$\mathscr{A} = \operatorname{End}^{\bullet}_{\mathrm{D}^b(\mathcal{O}G)}(\mathcal{O}[G/U] \oplus \mathbf{R\Gamma}_c(\mathbf{Y}, \mathcal{O})).$$

This algebra is an A_∞-algebra (that is to say, an algebra equipped with *higher products* $m_n \colon \mathscr{A}^{\otimes n} \to \mathscr{A}$ verifying certain compatibility conditions). In his thesis, Gonard [Go, §4.2] gives a complete description of this A_∞-algebra. For a review of A_∞-algebras, the reader is referred to [Go, §4.1].

Theorem 8.3.9 (Gonard). *If $\ell = 2$, then the derived category $\mathrm{D}^b(\mathcal{O}G)$ is equivalent to the derived category of the A_∞-algebra $\mathscr{A}$. This A_∞-algebra satisfies*

$$\dim_{\mathcal{O}} \mathscr{A} = 4(q+1) \quad \textit{and} \quad m_n = 0 \textit{ if } n \geqslant 4.$$

8.4. Alvis-Curtis Duality*

I would like to thank Marc Cabanes for suggesting this section to me, as well as for having provided a simple argument for the proof of the main result (Theorem 8.4.1).

> **Hypothesis.** *In this section we fix a commutative ring R in which p is invertible.*

REMARK – $\mathbb{Q}_\ell$, $\mathbb{Z}_\ell$, $\mathbb{F}_\ell$, as well as K, $\mathcal{O}$ and k are rings in which p is invertible. □

This hypothesis implies that the idempotent e_U can be viewed as an element of the group algebra RG. Recall that the (RG, RT)-bimodules $R[G/U]$ and RGe_U are isomorphic. Denote by

$$\begin{array}{rccc} \delta: & RGe_U \otimes_{RT} e_U RG & \longrightarrow & RG \\ & a \otimes_{RT} b & \longmapsto & ab. \end{array}$$

It is a morphism of (RG, RG)-bimodules. We set

$$\begin{array}{rccc} \delta^*: & RG & \longrightarrow & RGe_U \otimes_{RT} e_U RG \\ & a & \longmapsto & \displaystyle\sum_{g \in [G/B]} age_U \otimes_{RT} e_U g^{-1} \end{array}$$

the dual morphism. One may easily verify that δ^* is indeed a morphism of (RG, RG)-bimodules (indeed, it is enough to verify that $\delta^*(g) = g\delta^*(1) = \delta^*(1)g$ for all $g \in G$).

Denote by $\mathcal{D}$ the complex of (RG, RG)-bimodules

$$\mathcal{D} = \big(0 \longrightarrow RGe_U \otimes_{RT} e_U RG \xrightarrow{\ \delta\ } RG \longrightarrow 0\big)$$

and set

$$\mathcal{D}^* = \big(0 \longrightarrow RG \xrightarrow{\ \delta^*\ } RGe_U \otimes_{RT} e_U RG \longrightarrow 0\big),$$

the dual complex. Even though this will not influence the principal result of this section we suppose, in accordance with standard conventions, that the non-zero terms of $\mathcal{D}$ (respectively $\mathcal{D}^*$) occur in degree 0 and 1 (respectively -1 and 0). If M is a (RG, RG)-bimodule, we also use the notation M for the complex $0 \to M \to 0$, where M is in degree 0.

Theorem 8.4.1. *In the homotopy category $\mathsf{K}^b(RG, RG)$, we have*

$$\mathcal{D} \otimes_{RG} \mathcal{D}^* \simeq_{\mathsf{K}} \mathcal{D}^* \otimes_{RG} \mathcal{D} \simeq_{\mathsf{K}} RG.$$

Proof. To simplify notation set $e = e_U$ and $M = RGe \otimes_{RT} eRG$. Furthermore, we will write $\otimes$ for the tensor product $\otimes_{RT}$. We have

$$\mathscr{D} \otimes_{RG} \mathscr{D}^* = \big(0 \longrightarrow M \xrightarrow{f} RG \oplus (RGe \otimes eRGe \otimes eRG) \xrightarrow{g} M \longrightarrow 0\big),$$

where

$$f(a \otimes b) = ab \oplus \Big(\sum_{g \in [G/B]} a \otimes bge \otimes eg^{-1}\Big)$$

and

$$g(x \oplus (a \otimes b \otimes c)) = \delta^*(x) - ab \otimes c.$$

The endomorphism $\mathscr{F}$ of the RG-module RGe defined by $\mathscr{F}(x) = qxese$ is an isomorphism. Indeed, the calculation completed over $\mathbb{Z}_\ell$ in Theorem 7.2.4 remains valid here and shows that

$$\mathscr{F} \circ (\mathscr{F} - E) = q\mathbf{d}(-1).$$

We will need the following lemma.

Lemma 8.4.2. *The morphism*

$$\begin{array}{ccc} M \oplus M & \longrightarrow & RGe \otimes eRGe \otimes eRG \\ (x \otimes x') \oplus (y \otimes y') & \longmapsto & x \otimes e \otimes x' + \mathscr{F}^{-1}(y) \otimes ese \otimes y' \end{array}$$

is an isomorphism of (RG, RG)*-bimodules.*

Proof (of Lemma 8.4.2). By Bruhat decomposition 1.1.4, we have $eRGe = eRBe \oplus eR[BsB]e$ (as (RT, RT)-bimodules). As a consequence,

$$RGe \otimes eRGe = (RGe \otimes eRBe) \oplus (RGe \otimes eR[BsB]e).$$

It is enough to show that the morphisms

$$\begin{array}{ccc} RGe & \longrightarrow & RGe \otimes eRBe \\ x & \longmapsto & x \otimes e \end{array} \quad \text{and} \quad \begin{array}{ccc} RGe & \longrightarrow & RGe \otimes eR[BsB]e \\ y & \longmapsto & \mathscr{F}^{-1}(y) \otimes ese \end{array}$$

are isomorphisms of (RG, RT)-bimodules. This is a straightforward consequence of the fact that, in both cases, the inverse is given by the formula $a \otimes b \mapsto ab$. □

Using the isomorphism of Lemma 8.4.2, the complex $\mathscr{D} \otimes_{RG} \mathscr{D}^*$ is isomorphic (in the category $C^b(RG, RG)$) to the complex

$$0 \longrightarrow M \xrightarrow{f'} RG \oplus M \oplus M \xrightarrow{g'} M \longrightarrow 0,$$

where

$$f'(m) = \delta(m) \oplus m \oplus \delta^*(\delta(m)) - m$$

and
$$g'(a\oplus m\oplus m') = \delta^*(a) - m - m'.$$

This uses the fact that $\delta^*\delta(a\otimes b) = a\otimes b + qase\otimes es^{-1}b$.

Now consider

$$\begin{aligned} \varphi\colon\ RG\oplus M\oplus M &\longrightarrow \qquad\qquad RG\oplus M\oplus M \\ a\oplus m\oplus m' &\longmapsto a-\delta(m)\oplus m\oplus(\delta^*(a)-m-m'). \end{aligned}$$

Then φ is an isomorphism of (RG, RG)-bimodules, which shows that the complex $\mathscr{D}\otimes_{RG}\mathscr{D}^*$ is isomorphic (in the category $\mathsf{C}^b(RG, RG)$) to the complex

$$0 \longrightarrow M \xrightarrow{f''} RG\oplus M\oplus M \xrightarrow{g''} M \longrightarrow 0,$$

where $f''(m) = 0\oplus m\oplus 0$ and $g''(a\oplus m\oplus m') = m'$. In other words,

$$\mathscr{D}\otimes_{RG}\mathscr{D}^* \simeq_{\mathsf{C}} RG\oplus\mathscr{C}\oplus\mathscr{C}[1],$$

where $\mathscr{C}$ is the complex $0\to M \xrightarrow{\mathrm{Id}_M} M \to 0$ in which the non-zero terms are concentrated in degrees 0 and 1. As $\mathscr{C}$ (and therefore $\mathscr{C}[1]$) is homotopic to zero, we deduce that $\mathscr{D}\otimes_{RG}\mathscr{D}^*$ is homotopic to RG.

The fact that $\mathscr{D}^*\otimes_{RG}\mathscr{D}$ is also homotopic to RG is shown in exactly the same fashion. □

Corollary 8.4.3. *The functor $\mathscr{D}\otimes_{RG} -$ induces an auto-equivalence of the homotopy category $\mathsf{K}^b(RG, RG)$ (as well as of the derived category $\mathsf{D}^b(RG, RG)$).*

Exercises

8.1. If $\alpha\in T^\wedge$ (respectively $\theta\in T'^\wedge$), show that $\mathscr{O}Tb_\alpha \simeq \mathscr{O}S_\ell$ (respectively $\mathscr{O}T'b'_\theta \simeq \mathscr{O}S'_\ell$).

Now suppose that ℓ is odd. Show that the $\mathscr{O}$-algebras $\mathscr{O}Nb_1$ and $\mathscr{O}Nb_{\alpha_0}$ are isomorphic, and similarly for the $\mathscr{O}$-algebras $\mathscr{O}N'b'_1$ and $\mathscr{O}N'b'_{\theta_0}$.

8.2. Show that, if ℓ divides $|G|/(q-1)$, then the algebras $\mathrm{End}_{\mathbb{Z}_\ell G}\mathbb{Z}_\ell[G/U]$ and $\mathbb{Z}_\ell N$ are not isomorphic.

Show that, if ℓ divides $|G|/(q+1)$, then the algebras $\mathrm{End}_{\mathsf{D}^b(\mathbb{Z}_\ell G)}\mathbf{R\Gamma}_c(\mathbf{Y},\mathbb{Z}_\ell)$ and $\mathbb{Z}_\ell N'$ are not isomorphic.

8.3* If R is a field of characteristic different from p, the equivalence of homotopy categories $\mathscr{D}\otimes_{RG} -$ induces an isomorphism between the Grothendieck groups that we will denote $\mathrm{D}_R\colon \mathscr{K}_0(RG) \xrightarrow{\sim} \mathscr{K}_0(RG)$. Show that

$$\mathrm{D}_R[M] = [\mathscr{R}_R({}^*\mathscr{R}_R M)] - [M].$$

Calculate D_K.

Chapter 9
Unequal Characteristic: Simple Modules, Decomposition Matrices

Hypothesis. *In this chapter, as in the two previous chapters, we suppose that ℓ is a prime number different from p.*

In this chapter we will determine the simple kG-modules and the decomposition matrix $\mathrm{Dec}(\mathcal{O}G)$, as a function of the prime number ℓ. This study will be carried out block by block, or more precisely by type of block (nilpotent, quasi-isolated, principal). When the defect group is cyclic we will also give the Brauer tree. We refer the reader to Appendix B for the definitions of these concepts.

To carry out this study, we will use the equivalences of categories constructed in the previous chapter. When these equivalences are Morita equivalences the simple modules correspond to one another and the decomposition matrices are preserved (as is the Brauer tree). When the equivalences in question are genuine Rickard equivalences (which only occurs for the principal block when ℓ is odd and divides $q+1$) the only property that is conserved is the number of simple modules.

9.1. Preliminaries

9.1.1. Induction and Decomposition Matrices

By extension of scalars from to K to k, Harish-Chandra induction $\mathscr{R}$ (respectively Deligne-Lusztig induction $\mathscr{R}'$) induces two functors $\mathscr{R}_K$ and $\mathscr{R}_k$ (respectively $\mathscr{R}'_K$ and $\mathscr{R}'_k$). These functors induce linear maps, denoted R and R_k (respectively R' and R'_k) between the Grothendieck groups of T (re-

C. Bonnafé, *Representations of* $\mathrm{SL}_2(\mathbb{F}_q)$, Algebra and Applications 13,
DOI 10.1007/978-0-85729-157-8_9,

spectively T') and G:

$$\mathrm{R}\colon \mathscr{K}_0(KT) \longrightarrow \mathscr{K}_0(KG),$$
$$\mathrm{R}_k\colon \mathscr{K}_0(kT) \longrightarrow \mathscr{K}_0(kG),$$
$$\mathrm{R}'\colon \mathscr{K}_0(KT') \longrightarrow \mathscr{K}_0(KG)$$

and

$$\mathrm{R}'_k\colon \mathscr{K}_0(kT') \longrightarrow \mathscr{K}_0(kG).$$

Because the functor $\mathscr{R}$ (respectively $\mathscr{R}'$) is induced by an $\mathscr{O}G$-module which is $\mathscr{O}$-free (respectively a complex of $\mathscr{O}G$-modules which are $\mathscr{O}$-free), the diagrams

(9.1.1)
$$\begin{array}{ccc} \mathscr{K}_0(KT) & \xrightarrow{\quad \mathrm{R} \quad} & \mathscr{K}_0(KG) \\ {\scriptstyle \mathrm{dec}_{\mathscr{O}T}}\downarrow & & \downarrow{\scriptstyle \mathrm{dec}_{\mathscr{O}G}} \\ \mathscr{K}_0(kT) & \xrightarrow{\quad \mathrm{R}_k \quad} & \mathscr{K}_0(kG) \end{array}$$

and

(9.1.2)
$$\begin{array}{ccc} \mathscr{K}_0(KT') & \xrightarrow{\quad \mathrm{R}' \quad} & \mathscr{K}_0(KG) \\ {\scriptstyle \mathrm{dec}_{\mathscr{O}T'}}\downarrow & & \downarrow{\scriptstyle \mathrm{dec}_{\mathscr{O}G}} \\ \mathscr{K}_0(kT') & \xrightarrow{\quad \mathrm{R}'_k \quad} & \mathscr{K}_0(kG) \end{array}$$

are commutative.

9.1.2. Dimensions of Modules and Restriction to U

In 5.2.1 we introduced a non-trivial linear character χ_+ of the additive group $\mathbb{F}_q^+$. We denote by $\chi_k\colon \mathbb{F}_q^+ \to k^\times$ the composition $\mathbb{F}_q^+ \xrightarrow{\chi_+} \mathscr{O}^\times \twoheadrightarrow k^\times$. If $x \in \mathbb{F}_q$ we denote by ψ_x the linear character of U (with values in $k^\times$) defined by

$$\psi_x(\mathbf{u}(z)) = \chi_k(xz)$$

for all $z \in \mathbb{F}_q$. We set

$$e_x = \frac{1}{q}\sum_{u\in U} \psi_x(u^{-1})u \in kU.$$

If $a \in \mathbb{F}_q^\times$, then

$$\mathbf{d}(a) e_x \mathbf{d}(a^{-1}) = e_{a^{-2}x} \tag{9.1.3}$$

(see Proposition 1.1.2(b)). Consequently, if M is a kG-module, then

$$\mathbf{d}(a) e_x M = e_{a^{-2}x} M. \tag{9.1.4}$$

On the other hand, as the order of U is invertible in k, we have

$$e_0 M = M^U \qquad \text{and} \qquad M = \bigoplus_{x \in \mathbb{F}_q} e_x M. \tag{9.1.5}$$

The next proposition follows from these observations.

Proposition 9.1.6. *Let M be a kG-module and let x and y be two elements of $\mathbb{F}_q^\times$ such that xy is not a square. Then*

$$\dim_k M = \dim_k M^U + \frac{q-1}{2}(\dim_k e_x M + \dim_k e_y M).$$

In particular,

$$\dim_k M \equiv \dim_k M^U \mod \frac{q-1}{2}. \tag{9.1.7}$$

9.2. Nilpotent Blocks

Let α (respectively θ) be a linear character of $T_{\ell'}$ (respectively $T'_{\ell'}$) such that $\alpha^2 \neq 1$ (respectively $\theta^2 \neq 1$). By Proposition 8.1.1 (respectively 8.1.4), the bimodule $\mathcal{O}[G/U]b_\alpha$ (respectively $H_c^1(\mathbf{Y}, \mathcal{O})b'_\theta$) induces a Morita equivalence between A_α and $\mathcal{O}Tb_\alpha$ (respectively between A'_θ and $\mathcal{O}T'b'_\theta$). By extension of scalars we deduce that tensoring with $k[G/U]b_\alpha$ (respectively $H_c^1(\mathbf{Y}, k)b'_\theta$) induces a Morita equivalence between kA_α and kTb_α (respectively kA'_θ and $kT'b'_\theta$). Now, kTb_α (respectively $kT'b'_\theta$) has only one simple module k_α (respectively k_θ), that is to say the k-vector space k on which T (respectively T') acts via the linear character $T \xrightarrow{\alpha} \mathcal{O}^\times \twoheadrightarrow k^\times$ (respectively $T' \xrightarrow{\theta} \mathcal{O}^\times \twoheadrightarrow k^\times$).

As a consequence,

$$\operatorname{Irr} kA_\alpha = \{\mathcal{R}_k k_\alpha\} = \{\operatorname{Ind}_B^G k_{\tilde{\alpha}}\}$$

and

$$\operatorname{Irr} kA'_\theta = \{H_c^1(\mathbf{Y}, k) \otimes_{kT'} k_\theta\}.$$

Note that $|\mathrm{Irr}\, kA_\alpha| = |\mathrm{Irr}\, kA'_\theta| = 1$, as should be the case for any nilpotent block. The decomposition matrices are given by

$$\mathrm{Dec}(A_\alpha) = \begin{pmatrix} 1 \\ 1 \\ \vdots \\ 1 \end{pmatrix} \quad \text{and} \quad \mathrm{Dec}(A'_\theta) = \begin{pmatrix} 1 \\ 1 \\ \vdots \\ 1 \end{pmatrix},$$

where the number of lines of $\mathrm{Dec}(A_\alpha)$ (respectively $\mathrm{Dec}(A'_\theta)$) is $|S_\ell|$ (respectively $|S'_\ell|$) because

$$\mathrm{Irr}\, KA_\alpha = \{\mathrm{R}(\alpha\eta) \mid \eta \in S_\ell^\wedge\}$$

and

$$\mathrm{Irr}\, KA'_\theta = \{\mathrm{R}'(\theta\eta) \mid \eta \in S_\ell'^\wedge\}.$$

***Brauer trees**.** The only non-exceptional character of A_α (respectively A'_θ) is $\mathrm{R}(\alpha)$ (respectively $\mathrm{R}'(\theta)$). The Brauer trees are therefore given by

$\mathscr{T}_{A_\alpha}$ 

and

$\mathscr{T}_{A'_\theta}$ .

9.3. Quasi-Isolated Blocks

Hypothesis. *In this and only this section we suppose that ℓ is odd (and different from p).*

We recall that this hypothesis is necessary for the existence of quasi-isolated blocks.

First, by Proposition 8.2.1 (respectively 8.2.2), the $(A_{\alpha_0}, \mathcal{O}Tb_{\alpha_0})$-bimodule $\mathcal{O}[G/U]b_{\alpha_0}$ (respectively the $(A'_{\theta_0}, \mathcal{O}T'b'_{\theta_0})$-bimodule $H_c^1(\mathbf{Y}, \mathcal{O})b'_{\theta_0}$) can be extended to an $(A_{\alpha_0}, \mathcal{O}Nb_{\alpha_0})$-bimodule (respectively to an $(A'_{\theta_0}, \mathcal{O}N'b'_{\theta_0})$-bimodule), and the latter induces a Morita equivalence. The action of the element s of N (respectively s' of N') is given by $\mathscr{F}/\sqrt{q}$ (respectively $\mathscr{F}'/\sqrt{-q}$).

Now, the k-algebra kNb_{α_0} (respectively $kN'b'_{\theta_0}$) admits two simple modules $k^\pm_{\alpha_0}$ (respectively $k^\pm_{\theta_0}$) associated to the linear characters $\chi^\pm_{\alpha_0}$ (respectively $\chi'^\pm_{\theta_0}$) of N (respectively N') defined in Section 6.2. Via the Morita equivalence they correspond to two simple modules of kA_{α_0} (respectively kA'_{θ_0}) that we denote by $\mathscr{R}_{\pm}(\alpha_0)$ (respectively $\mathscr{R}'_{\pm}(\theta_0)$). By 9.1.1 (respectively 9.1.2), these are the reductions modulo $\mathfrak{l}$ of $\mathscr{O}$-free $\mathscr{O}G$-modules admitting $\mathrm{R}_{\pm}(\alpha_0)$ (respectively $\mathrm{R}'_{\pm}(\theta_0)$) as characters. As

$$\operatorname{Irr} KA_{\alpha_0} = \{\mathrm{R}_+(\alpha_0), \mathrm{R}_-(\alpha_0)\} \,\dot{\cup}\, \{\mathrm{R}(\alpha_0\eta) \mid \eta \in [S_\ell^\wedge/\equiv], \eta \neq 1\}$$

and $\qquad \operatorname{Irr} KA'_{\theta_0} = \{\mathrm{R}'_+(\theta_0), \mathrm{R}'_-(\theta_0)\} \,\dot{\cup}\, \{\mathrm{R}(\theta_0\eta) \mid \eta \in [S_\ell'^\wedge/\equiv], \eta \neq 1\},$

we deduce that

$$\operatorname{Irr} kA_{\alpha_0} = \{\mathscr{R}_+(\alpha_0), \mathscr{R}_-(\alpha_0)\},$$

that

$$\operatorname{Irr} kA'_{\theta_0} = \{\mathscr{R}'_+(\theta_0), \mathscr{R}'_-(\theta_0)\}$$

and that the decomposition matrices are given by

$$\operatorname{Dec}(A_{\alpha_0}) = \begin{pmatrix} 1 & 0 \\ 0 & 1 \\ 1 & 1 \\ \vdots & \vdots \\ 1 & 1 \end{pmatrix} \quad \text{and} \quad \operatorname{Dec}(A'_{\theta_0}) = \begin{pmatrix} 1 & 0 \\ 0 & 1 \\ 1 & 1 \\ \vdots & \vdots \\ 1 & 1 \end{pmatrix}.$$

Here, the first two lines of $\operatorname{Dec}(A_{\alpha_0})$ (respectively $\operatorname{Dec}(A'_{\theta_0})$) correspond to the characters $\mathrm{R}_{\pm}(\alpha_0)$ (respectively $\mathrm{R}'_{\pm}(\theta_0)$). The number of lines of $\operatorname{Dec}(A_{\alpha_0})$ (respectively $\operatorname{Dec}(A'_{\theta_0})$) is equal to $(|S_\ell|+3)/2$ (respectively $(|S'_\ell|+3)/2$).

Brauer trees*. The non-exceptional characters of A_{α_0} (respectively A'_{θ_0}) are $\mathrm{R}_{\pm}(\alpha_0)$ (respectively $\mathrm{R}'_{\pm}(\theta_0)$). Therefore, the Brauer trees are given by

$\mathscr{T}_{A_{\alpha_0}}$

and

$\mathscr{T}_{A'_{\theta_0}}$

.

9.4. The Principal Block

9.4.1. Preliminaries

We identify $\mathscr{R}_k k$ with $k[\mathbf{P}^1(\mathbb{F}_q)]$ and, via this identification, we denote by St_G^k the k-vector subspace of $k[\mathbf{P}^1(\mathbb{F}_q)]$ equal to

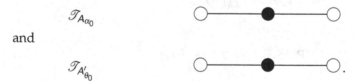

$$\mathrm{St}_G^k = \{ \sum_{l \in \mathbf{P}^1(\mathbb{F}_q)} \kappa_l l \mid \sum_{l \in \mathbf{P}^1(\mathbb{F}_q)} \kappa_l = 0 \}.$$

We denote by v the element

$$v = \sum_{l \in \mathbf{P}^1(\mathbb{F}_q)} l \in k[\mathbf{P}^1(\mathbb{F}_q)].$$

Then kv and St_G^k are kG-submodules of $k[\mathbf{P}^1(\mathbb{F}_q)]$, with kv being isomorphic to the trivial module. We have

(9.4.1) $$\mathrm{dec}_{\mathcal{O}G}\, \mathrm{St}_G = [\,\mathrm{St}_G^k\,]_{kG}.$$

Moreover, as $|\mathbf{P}^1(\mathbb{F}_q)| = q+1$, we deduce that

(9.4.2) $kv \subseteq \mathrm{St}_G^k$ *if and only if* ℓ *divides* $q+1$.

We finish with the following result.

Proposition 9.4.3. *If* $\ell \neq 2$, *then* kv *and* St_G^k *are the only two non-trivial* kG-*submodules of* $k[\mathbf{P}^1(\mathbb{F}_q)]$.

Proof. We begin with some notation. If $z \in \mathbb{F}_q$, we set $l_z = [z;1] \in \mathbf{P}^1(\mathbb{F}_q)$ and $l_\infty = [1;0]$. We then have

$$s \cdot l_z = l_{-z^{-1}} \qquad \text{and} \qquad \mathbf{u}(x) \cdot l_z = l_{z+x}$$

for all $x \in \mathbb{F}_q$. Note that, with the usual conventions, the two previous formulas make sense and are valid even when $z \in \{0, \infty\}$.

If $x \in \mathbb{F}_q$, we set

$$v_x = \sum_{z \in \mathbb{F}_q} \chi_k(-xz) l_z.$$

Then

$$k[\mathbf{P}^1(\mathbb{F}_q)] = kv \oplus \Big(\bigoplus_{x \in \mathbb{F}_q} kv_x \Big).$$

The action of the subgroup U is entirely determined by the following formulas:

(∗) $$e_0 \cdot v = v \qquad \text{and} \qquad e_x \cdot v_y = \delta_{x=y} v_y$$

for all $x, y \in \mathbb{F}_q$.

Let M be a non-trivial kG-submodule of $k[\mathbf{P}^1(\mathbb{F}_q)]$. If U acts trivially on M, then G acts trivially on M because G is generated by the conjugates of U (see Lemma 1.2.2). In this case, $M = kv$ and the proposition holds. We may therefore suppose that U (i.e. G) does not act trivially on M. As a consequence, there exists an element $x \in \mathbb{F}_q^\times$ such that

(1) $$e_x M \neq 0.$$

In other words, $v_x \in M$ by (∗). Again by (∗), we have

$$e_y s \cdot v_x = e_y l_\infty + \frac{1}{q}\Big(\sum_{z \in \mathbb{F}_q^\times} \chi_k\big(z + \frac{xy}{z}\big)\Big) v_y.$$

But, by Exercise 9.1 (which is easy) and the fact that $\ell \neq 2$, there exists $y \in \mathbb{F}_q^\times$ such that xy is not a square in $\mathbb{F}_q$ and

$$\sum_{z \in \mathbb{F}_q^\times} \chi_k\big(z + \frac{xy}{z}\big) \neq 0.$$

As a consequence,

(2) $$e_y M \neq 0.$$

On the other hand, the injective morphism of kG-modules $M \hookrightarrow \mathrm{Ind}_B^G k$, which, by adjunction, furnishes a non-trivial morphism of kB-modules $\mathrm{Res}_B^G M \to k$. Hence,

(3) $$e_0 M \neq 0.$$

It then follows from (1), (2), (3) and Proposition 9.1.6 that $\dim_k M \geqslant q$.

Now set $M' = M \cap \mathrm{St}_G^k$. Then $\dim_k M' \geqslant q - 1 \geqslant 2$ and again M' is a kG-submodule of $k[\mathbf{P}^1(\mathbb{F}_q)]$. As its dimension is at least 2, G does not act trivially on M' and the previous argument applies again to M'. We then obtain that $\dim_k M' \geqslant q$ and, as $M' \subseteq \mathrm{St}_G^k$, we conclude that $M' = \mathrm{St}_G^k$, that is to say $\mathrm{St}_G^k \subseteq M$. □

9.4.2. *The Case when* ℓ *is Odd and Divides* $q-1$

Hypothesis. *In this and only this subsection we assume that* ℓ *is* ***odd*** *and* ***divides*** $q-1$.

This case may be studied in a similar manner to the quasi-isolated block A_{α_0}. As ℓ does not divide $q+1$, we deduce from 9.4.2 that

$$k[G/B] = \mathrm{St}_G^k \oplus kv.$$

Hence the two simple modules of kA_1 which correspond to the two simple kN-modules k_1 and k_ε via the Morita equivalence are kv and St_G^k:

$$\mathrm{Irr}\, kA_1 = \{k, \mathrm{St}_G^k\}.$$

As

$$\mathrm{Irr}\, KA_1 = \{1_G, \mathrm{St}_G\} \,\dot{\cup}\, \{\mathrm{R}(\eta) \mid \eta \in [S_\ell^\wedge/\equiv], \eta \neq 1\},$$

we conclude that the decomposition matrix is

$$\mathrm{Dec}(A_1) = \begin{pmatrix} 1 & 0 \\ 0 & 1 \\ 1 & 1 \\ \vdots & \vdots \\ 1 & 1 \end{pmatrix}.$$

Here, the two first lines are indexed by 1_G and St_G and the number of lines is $(|S_\ell|+1)/2$.

***Brauer trees**.** The non-exceptional characters of A_1 are 1_G and St_G, and the Brauer tree is given by

$\mathscr{T}_{A_1}$ .

9.4.3. *The Case when ℓ is Odd and Divides $q+1$*

Hypothesis. *In this and only this subsection we suppose that ℓ is odd and divides $q+1$.*

In this case it follows from 9.4.2 that St_G^k is not simple. Indeed, it contains kv as a submodule. Let us denote $\overline{\mathrm{St}}_G^k = \mathrm{St}_G^k/kv$.

Proposition 9.4.4. *The kG-module $k[G/B]$ is projective and indecomposable. The kG-module $\overline{\mathrm{St}}_G^k$ is simple.*

Proof. As ℓ is odd, the simplicity of $\overline{\mathrm{St}}_G^k$ follows from Proposition 9.4.3. The projectivity of $k[G/B]$ follows because ℓ does not divide the cardinality of B. The fact that $k[G/B]$ is indecomposable follows from Proposition 9.4.3 and the fact that $kv \subseteq \mathrm{St}_G^k$. □

As a consequence,

(9.4.5) $$\operatorname{Irr} kA_1 = \{k, \overline{\mathrm{St}}_G^k\}.$$

As

$$\operatorname{Irr} KA_1 = \{1_G, \mathrm{St}_G\} \,\dot\cup\, \{\mathrm{R}(\eta) \mid \eta \in [S_\ell^\wedge/\equiv], \eta \neq 1\},$$

we conclude that the decomposition matrix has the form

$$\mathrm{Dec}(A_1) = \begin{pmatrix} 1 & 0 \\ 1 & 1 \\ 0 & 1 \\ \vdots & \vdots \\ 0 & 1 \end{pmatrix}.$$

Brauer trees*. The non-exceptional characters of A_1 are 1_G and St_G and the Brauer tree is given by

$\mathscr{T}_{A_1}$

9.4.4. *The Case when* $\ell = 2$

> **Hypothesis.** *In this and only this subsection we suppose that* $\ell = 2$.

Fix an $\mathscr{O}$-free $\mathscr{O}G$-module $V_\pm$ such that $KV_\pm$ admits the character $\mathrm{R}'_\pm(\theta_0)$. Set

$$\overline{\mathrm{St}}_\pm^k = kV_\pm.$$

Then

(9.4.6) $$\dim_k \overline{\mathrm{St}}_\pm^k = \frac{q-1}{2}$$

and, by Proposition 5.2.7,

(9.4.7) $$\dim_k e_U \overline{\mathrm{St}}_\pm^k = 0.$$

In order to show the last equality one uses the fact that $|U|$ is invertible in k. It then follows from Proposition 9.1.6 that

(9.4.8) $\overline{\mathrm{St}}^k_\pm$ *is a simple kG-module.*

Proposition 9.4.9. *If* $\ell = 2$, *then* $\mathrm{Irr}\, kA_1 = \{k, \overline{\mathrm{St}}^k_+, \overline{\mathrm{St}}^k_-\}$. *Moreover,* $\overline{\mathrm{St}}^k_+ \not\simeq \overline{\mathrm{St}}^k_-$.

Proof. The fact that $\overline{\mathrm{St}}^k_+ \not\simeq \overline{\mathrm{St}}^k_-$ follows immediately from considering the restriction to U and Proposition 5.2.7.

On the other hand, recall that $\mathrm{Irr}\, KA_1$ is the set of irreducible characters of $\mathrm{R}(\alpha)$ (for $\alpha \in S_2^\wedge$) and $\mathrm{R}'(\theta)$ (for $\theta \in S_2'^\wedge$). But, for such characters, it follows from 9.1.1 and 9.1.2 that

$$\mathrm{dec}_{\mathcal{O}G}\, \mathrm{R}(\alpha) = \mathrm{dec}_{\mathcal{O}G}\, \mathrm{R}(1) = [k]_{kG} + \mathrm{dec}_{\mathcal{O}G}\, \mathrm{St}_G$$

and
$$\mathrm{dec}_{\mathcal{O}G}\, \mathrm{R}'(\theta) = \mathrm{dec}_{\mathcal{O}G}\, \mathrm{R}'(1) = -[k]_{kG} + \mathrm{dec}_{\mathcal{O}G}\, \mathrm{St}_G.$$

As
$$\mathrm{dec}_{\mathcal{O}G}\, \mathrm{R}'(\theta_0) = [\overline{\mathrm{St}}^k_+]_{kG} + [\overline{\mathrm{St}}^k_-]_{kG},$$

the proposition then follows from 9.4.8 and the fact that the isomorphism class of every simple kA_1-module appears in at least one $\mathrm{dec}_{\mathcal{O}G}\, \chi$, where $\chi \in \mathrm{Irr}\, KA_1$ (because $\mathrm{dec}_{\mathcal{O}G}\, [KA_1]_{KG} = [kA_1]_{kG}$). □

The calculations completed in the proof of the previous proposition allow us to compute the decomposition matrix of A_1 (using also Proposition 5.2.7), which is given in Table 9.1 below. In this table, $m = (|S_2| - 2)/2$ and $m' = (|S_2'| - 2)/2$ and $\alpha_1, \ldots, \alpha_m$ (respectively $\theta_1, \ldots, \theta_{m'}$) give a set of representatives for $(S_2^\wedge \setminus \{1, \alpha_0\})/\equiv$ (respectively $(S_2'^\wedge \setminus \{1, \theta_0\})/\equiv$). Note that, if $q \equiv 1 \mod 4$ (respectively $q \equiv 3 \mod 4$), then $m' = 0$ (respectively $m = 0$).

REMARK 9.4.10 – Note that there is no Brauer tree for A_1 because the defect group of A_1 is a Sylow 2-subgroup of G and is therefore not cyclic (see Theorem 1.4.3). □

REMARK 9.4.11 – When $\ell = 2$, the principal k-block of $\widetilde{G} = \mathrm{GL}_2(\mathbb{F}_q)$ has only two simple modules, contrarily to the principal k-block of $G = \mathrm{SL}_2(\mathbb{F}_q)$. □

Exercises

9.1 (Katz). [†] *In this and only this exercise we suppose that* $\ell \neq 2$. If $a \in \mathbb{F}_q$, set

$$\mathrm{Kl}(a) = \sum_{z \in \mathbb{F}_q^\times} \chi_k\Big(z + \frac{a}{z}\Big).$$

[†] I would like to thank N. Katz for suggesting this exercise to me.

Table 9.1 Decomposition matrix for A_1 when $\ell = 2$.

	k	$\overline{\mathrm{St}}_+^k$	$\overline{\mathrm{St}}_-^k$
1_G	1	0	0
St_G	1	1	1
$R'_+(\theta_0)$	0	1	0
$R'_-(\theta_0)$	0	0	1
$R_+(\alpha_0)$	1	1	0
$R_+(\alpha_0)$	1	0	1
$R(\alpha_1)$	2	1	1
$\vdots$	$\vdots$	$\vdots$	$\vdots$
$R(\alpha_m)$	2	1	1

$\mathrm{Dec}(A_1)$ when $q \equiv 1 \mod 4$

	k	$\overline{\mathrm{St}}_+^k$	$\overline{\mathrm{St}}_-^k$
1_G	1	0	0
St_G	1	1	1
$R'_+(\theta_0)$	0	1	0
$R'_-(\theta_0)$	0	0	1
$R_+(\alpha_0)$	1	1	0
$R_+(\alpha_0)$	1	0	1
$R'(\theta_1)$	0	1	1
$\vdots$	$\vdots$	$\vdots$	$\vdots$
$R'(\theta_{m'})$	0	1	1

$\mathrm{Dec}(A_1)$ when $q \equiv 3 \mod 4$

The purpose of this exercise is to show that there exists an element a of $\mathbb{F}_q^\times$ which is not a square and such that $\mathrm{Kl}(a) \neq 0$ (this fact is used in the proof of Proposition 9.4.3). The value $\mathrm{Kl}(a)$ is called a *Kloosterman sum*. The following exercise does not make use of the finer properties of such sums.

(a) Show that, if $a \in \mathbb{F}_q$ and $b \in \mathbb{F}_q^\times$, then $\mathrm{Kl}(ab^2) = \sum_{z \in \mathbb{F}_q^\times} \chi_k(b(z + \frac{a}{z}))$.

(b) Let a, z and z' be three elements of $\mathbb{F}_q^\times$ such that $z + \dfrac{a}{z} = z' + \dfrac{a}{z'}$. Show that $z = z'$ or $zz' = a$.

(c) From now on fix an element a_0 of $\mathbb{F}_q$ which is not a square. Show that the fibres of the map $\omega\colon \mathbb{F}_q^\times \to \mathbb{F}_q$, $z \mapsto z + \dfrac{a_0}{z}$ are of cardinality 0 or 2.

(d) If $\theta \in \mathbb{F}_q$ we set

$$f(\theta) = \sum_{b \in \mathbb{F}_q^\times} \chi_k(-b\theta)\,\mathrm{Kl}(a_0 b^2).$$

Using (a), show that

$$f(\theta) = \begin{cases} 1 - q & \text{if } |\omega^{-1}(\theta)| = 0, \\ q + 1 & \text{if } |\omega^{-1}(\theta)| = 2. \end{cases}$$

(e) Using (c) and the fact that $q-1$ and $q+1$ cannot both be zero as $\ell \neq 2$, show that there exists an element a of $\mathbb{F}_q^\times$ which is not a square and such that $\mathsf{Kl}(a) \neq 0$.

9.2. *In this and only this exercise we suppose that* $\ell = 2$. Fix a non-square $a_0 \in \mathbb{F}_q^\times$ and consider $C_+ = \{b^2 \mid b \in \mathbb{F}_q^\times\}$ and $C_- = a_0 C_+$. Set

$$\mathrm{St}_\pm^k = kv \oplus \Big(\bigoplus_{x \in C_\pm} kv_x \Big).$$

Use the argument of the proof of Proposition 9.4.3 to show that $\mathrm{St}_\pm^k$ is a kG-submodule of St_G^k and that the only non-trivial kG-submodules of $k[\mathbf{P}^1(\mathbb{F}_q)]$ are kv, St_+^k, St_-^k and St_G^k.

Conclude that $k[G/B]$ is indecomposable and determine the structure of the successive quotients $\mathrm{Rad}^i(k[\mathbf{P}^1(\mathbb{F}_q)])/\mathrm{Rad}^{i+1}(k[\mathbf{P}^1(\mathbb{F}_q)])$, for $i \geqslant 0$.

9.3. *In this and only this exercise, we suppose that* ℓ *divides* $q+1$. Let $\mathscr{H}$ be the algebra of endomorphisms of $k[G/B]$. Show that $\mathscr{H} \simeq k[X]/\langle X^2 \rangle$, where X is an indeterminant. Deduce a new proof of the fact that $k[G/B]$ is indecomposable.

9.4* Calculate $\mathrm{D}_k : \mathscr{K}_0(kG) \xrightarrow{\sim} \mathscr{K}_0(kG)$ according to the values of ℓ. Recall that D_k is defined in Exercise 8.3. (***Hint:*** Use Exercise 8.3 and the fact that $\mathrm{dec}_{\mathscr{O}G} \circ \mathrm{D}_K = \mathrm{D}_k \circ \mathrm{dec}_{\mathscr{O}G}$.)

Chapter 10
Equal Characteristic

Hypothesis. *In this and only this chapter we suppose that $\ell = p$.*

In this chapter we study the representations of the group G in equal, or natural, characteristic. A significant part of this chapter will be dedicated to the construction of the simple kG-modules. This classical construction generalises to the case of finite reductive groups. It turns out that the simple kG-modules are the restrictions of simple "rational representations" of the algebraic group $\mathbf{G} = \mathrm{SL}_2(\mathbb{F})$. Having obtained this description, the determination of the decomposition matrices is straightforward.

We then determine the (very simple) partition into blocks as well as the Brauer correspondents. There is only one block with trivial defect (which corresponds to the Steinberg character St_G) and the two other $\mathcal{O}$-blocks (which both have U as their defect group — recall that the normaliser of U is B) are only distinguished by the action of the centre Z. As the group U is abelian, Broué's conjecture predicts an equivalence of derived categories between the $\mathcal{O}$-blocks of G and their Brauer correspondent. This result was shown by Okuyama [Oku1], [Oku2], for the principal block and by Yoshii [Yo] for the nonprincipal one (with defect group U) but the proof is too difficult to be included in this book.

To finish off, if $q = p$, then U is cyclic and we determine the Brauer trees of the two blocks with defect group U.

REMARK – The methods used in this chapter are totally different from those employed in the rest of this book. In particular, we use neither the geometry of the Drinfeld curve, nor any cohomology theory with coefficients modulo p. All methods are entirely algebraic. □

C. Bonnafé, *Representations of* $\mathrm{SL}_2(\mathbb{F}_q)$, Algebra and Applications 13, DOI 10.1007/978-0-85729-157-8_10, © Springer-Verlag London Limited 2011

Terminology, notation. In this and only this chapter we will denote by $\mathbf{G}$ the group $\mathrm{SL}_2(\mathbb{F})$ and $\mathbf{B}$ (respectively $\mathbf{T}$, respectively $\mathbf{U}$) the subgroup of $\mathbf{G}$ formed by upper triangular matrices (respectively diagonal, respectively upper unitriangular matrices). Note that

$$\mathbf{B} = \mathbf{T} \ltimes \mathbf{U}.$$

We will call a *rational representation* of $\mathbf{G}$ a morphism of algebraic groups $\mathbf{G} \to \mathrm{GL}_{\mathbb{F}}(V)$, where V is a finite dimensional $\mathbb{F}$-vector space. In this case we will often simply say that V is a *rational* $\mathbf{G}$*-module,* in which case the morphism is implicit. We denote by $\mathscr{K}_0(\mathbf{G})$ the Grothendieck group of the category of rational representations of $\mathbf{G}$.

We will denote by $\varepsilon\colon \mathbf{T} \to \mathbb{F}^\times$, $\mathrm{diag}(a, a^{-1}) \mapsto a$, so that $(\varepsilon^n)_{n\in\mathbb{Z}}$ is the set of **rational** representations of $\mathbf{T}$. Similarly, we identify $\mathscr{K}_0(\mathbf{T})$ with the group algebra $\mathbb{Z}[\mathbb{Z}]$ written exponentially: $\mathscr{K}_0(\mathbf{T}) = \oplus_{n\in\mathbb{Z}} \mathbb{Z}\varepsilon^n$. We will denote by $\mathrm{Car}_{\mathbf{G}}\colon \mathscr{K}_0(\mathbf{G}) \to \mathscr{K}_0(\mathbf{T})$ the morphism induced by restriction. If V is a rational $\mathbf{G}$-module, we denote by $[V]_{\mathbf{G}}$ its class in $\mathscr{K}_0(\mathbf{G})$ and we will write $\mathrm{Car}_{\mathbf{G}} V$ instead of $\mathrm{Car}_{\mathbf{G}} [V]_{\mathbf{G}}$ in order to simplify notation.

10.1. Simple Modules

10.1.1. Standard or Weyl Modules

Denote by V_2 the natural rational $\mathbf{G}$-module of dimension 2, that is to say, $V_2 = \mathbb{F}^2$ equipped with the standard action of $\mathbf{G} = \mathrm{SL}_2(\mathbb{F})$. If n is a natural number, we denote by $\Delta(n)$ the rational $\mathbf{G}$-module defined by

$$\Delta(n) = \mathrm{Sym}^n(V_2),$$

where $\mathrm{Sym}^n(V_2)$ denotes the n-th symmetric power of the vector space V_2. These modules are called *standard modules* or *Weyl modules*. It is easy to verify that

$$\dim_{\mathbb{F}} \Delta(n) = n+1 \tag{10.1.1}$$

and that

$$\Delta(0) \textit{ is the } \textbf{\textit{trivial}}\ \mathbf{G}\textit{-module.} \tag{10.1.2}$$

We denote by (x,y) the canonical basis of V_2, so that

$$\begin{pmatrix} a & b \\ c & d \end{pmatrix} \cdot x = ax + cy \quad \text{and} \quad \begin{pmatrix} a & b \\ c & d \end{pmatrix} \cdot y = bx + dy.$$

We have,

$$\Delta(n) = \bigoplus_{i=0}^{n} \mathbb{F} x^{n-i} y^i.$$

As $t \cdot x^{n-i}y^i = \varepsilon^{n-2i}(t)x^{n-i}y^i$, we obtain the following formula:

(10.1.3) $$\mathrm{Car}_{\mathbf{G}} \Delta(n) = \sum_{i=0}^{n} \varepsilon^{n-2i}.$$

We will use the following preliminary results in the construction of the simple rational **G**-modules.

Lemma 10.1.4. $\Delta(n)^{\mathbf{U}} = \mathbb{F}x^n$. *If moreover* $0 \leqslant n \leqslant q-1$, *then* $\Delta(n)^U = \mathbb{F}x^n$.

Proof. It is clear that $x^n \in \Delta(n)^{\mathbf{U}} \subseteq \Delta(n)^U$. To show that $\Delta(n)^{\mathbf{U}} \subseteq \mathbb{F}x^n$, it is sufficient, after replacing q by a power if necessary, to suppose that $0 \leqslant n \leqslant q-1$ and to show that $\Delta(n)^U \subseteq \mathbb{F}x^n$. This will also show the second assertion. Now let f be an element of $\Delta(n)^U$. Let us write

$$f = \sum_{i=0}^{n} \lambda_i x^{n-i} y^i.$$

Let $\xi \in \mathbb{F}_q$. Then $\mathbf{u}(\xi) \cdot f = f$. Now,

$$\mathbf{u}(\xi) \cdot f = \sum_{i=0}^{n} \lambda_i x^{n-i} (y + \xi x)^i = \sum_{j=0}^{n} \Big(\sum_{i=j}^{n} \binom{i}{j} \xi^{i-j} \lambda_i \Big) x^{n-j} y^j.$$

Comparing coefficients of x^n, we therefore have, for all $\xi \in \mathbb{F}_q$,

$$\sum_{i=0}^{n} \lambda_i \xi^i = \lambda_0.$$

As a consequence, the polynomial $\sum_{i=1}^{n} \lambda_i X^i$ admits q distinct roots and is of degree $\leqslant q-1$. It is therefore zero, which shows that $\lambda_1 = \cdots = \lambda_n = 0$, and hence $f \in \mathbb{F}x^n$. □

Corollary 10.1.5. *If* V *is a non-zero* **G**-*submodule of* $\Delta(n)$, *then* V *contains* x^n. *If* $0 \leqslant n \leqslant q-1$ *and if* V *is a non-zero* G-*submodule of* $\mathrm{Res}_G^{\mathbf{G}} \Delta(n)$, *then* V *contains* x^n.

Proof. After replacing q by a power, we may suppose that $0 \leqslant n \leqslant q-1$: it is then enough to prove the second assertion. Let V be a non-zero G-submodule of $\mathrm{Res}_G^{\mathbf{G}} \Delta(n)$. Then $V^U \neq 0$ because U is a p-group [CuRe, Theorem 5.24]. Therefore $x^n \in V^U \subseteq V$ by Lemma 10.1.4. □

To complete this subsection we will give another construction of $\Delta(n)$, using the algebra $\mathbb{F}[\mathbf{G}]$ of regular functions on $\mathbf{G}$. Denote by $\tilde{\varepsilon}\colon \mathbf{B} \to \mathbb{F}^\times$ the extension of ε to $\mathbf{B}$ which is trivial on $\mathbf{U}$. If $n \in \mathbb{Z}$, set

$$\Delta'(n) = \{f \in \mathbb{F}[\mathbf{G}] \mid \forall\, (g,b) \in \mathbf{G} \times \mathbf{B},\ f(gb) = \tilde{\varepsilon}(b)^{-n} f(g)\}.$$

Proposition 10.1.6. *If $\Delta'(n) \neq 0$, then $n \geqslant 0$. In this case, the $\mathbf{G}$-modules $\Delta(n)$ and $\Delta'(n)$ are isomorphic.*

Proof. Equip V_2 with the structure of a rational $(\mathbf{G},\mathbf{B})$-bimodule, letting $\mathbf{B}$ act on the right via $v \cdot b = \tilde{\varepsilon}(b)v$ for all $b \in \mathbf{B}$ and $v \in V_2$. We denote by V_2^* the dual of V_2, and let (x^*, y^*) denote the dual basis of (x,y). Then V_2^* is, by duality, also equipped with the structure of a rational $(\mathbf{G},\mathbf{B})$-bimodule. We identify the symmetric algebra $\mathrm{Sym}(V_2)$ with the algebra $\mathbb{F}[V_2^*]$ of regular (i.e. polynomial) functions on V_2^*. This algebra inherits the structure of a $(\mathbf{G},\mathbf{B})$-bimodule. Via this action we may describe $\Delta(n)$ as follows, when $n \geqslant 0$:

$$(*) \qquad \Delta(n) = \{f \in \mathrm{Sym}(V_2) \mid \forall\, b \in \mathbf{B},\ f \cdot b = \tilde{\varepsilon}(b)^n f\}.$$

Set

$$\begin{array}{rccc} \varphi\colon & \mathrm{Sym}(V_2) & \longrightarrow & \mathbb{F}[\mathbf{G}] \\ & f & \longmapsto & (g \mapsto f(g \cdot y^*)). \end{array}$$

If $g, h, \gamma \in \mathbf{G}$ and if $f \in \mathbb{F}[\mathbf{G}]$, we set

$$(g \cdot f \cdot h)(\gamma) = f(g^{-1}\gamma h^{-1}).$$

This equips $\mathbb{F}[\mathbf{G}]$ with the structure of a $(\mathbf{G},\mathbf{G})$-bimodule. One may easily verify that φ is a morphism of $(\mathbf{G},\mathbf{B})$-bimodules and that

$$(**) \qquad \Delta'(n) = \{f \in \mathbb{F}[\mathbf{G}] \mid \forall\, b \in \mathbf{B},\ f \cdot b = \tilde{\varepsilon}(b)^n f\}.$$

Set

$$\mathbb{F}[\mathbf{G}]^{\rho(\mathbf{U})} = \{f \in \mathbb{F}[\mathbf{G}] \mid \forall\, u \in \mathbf{U},\ f \cdot u = f\}.$$

We will need the following result.

Lemma 10.1.7. *φ is injective and its image is exactly $\mathbb{F}[\mathbf{G}]^{\rho(\mathbf{U})}$.*

Proof (of Lemma 10.1.7). Let $f \in \mathrm{Sym}(V_2)$ be such that $\varphi(f) = 0$. Then f is zero on the $\mathbf{G}$-orbit of y^*. This orbit is dense in V_2^*, therefore $f = 0$. This shows the injectivity φ.

On the other hand, the $(\mathbf{G},\mathbf{B})$-equivariance of φ implies that its image is contained in $\mathbb{F}[\mathbf{G}]^{\rho(\mathbf{U})}$. For the other implication, we choose $f \in \mathbb{F}[\mathbf{G}]^{\rho(\mathbf{U})}$ and must show that f is in the image of φ. We denote by $\hat{f}$ the map defined on $V_2^* \setminus \{0\}$ by $\hat{f}(v^*) = f(g)$ when $g \in \mathbf{G}$ is such that $v^* = g \cdot y^*$. As $f \in \mathbb{F}[\mathbf{G}]^{\rho(\mathbf{U})}$ and $\mathbf{U}$ is the stabiliser of y^* in $\mathbf{G}$, the function $\hat{f}$ is well-defined. We will show that it is a regular function on $V_2^* \setminus \{0\}$.

To this end, write $V_2^* \setminus \{0\} = \mathscr{U}_x \dot\cup \mathscr{U}_y$, where

$$\mathscr{U}_v = \{v^* \in V_2^* \mid v^*(v) \neq 0\}$$

for all $v \in V_2$. It is sufficient to show that the restriction of $\hat{f}$ to $\mathscr{U}_v$ is regular. Fix $g \in \mathbf{G}$ such that $g \cdot y = v$. Set $\mathbf{B}^- = \mathbf{T} \ltimes \mathbf{U}^-$, where $\mathbf{U}^-$ is the subgroup of $\mathbf{G}$ formed by the lower unitriangular matrices. Then $\mathbf{B}^-$ stabilises the line $\mathbb{F}y$ and the map

$$\begin{aligned} \rho_v \colon \mathbf{B}^- &\longrightarrow \mathscr{U}_v \\ b &\longmapsto gb \cdot y^* \end{aligned}$$

is clearly an isomorphism of varieties. Moreover, if $v^* \in \mathscr{U}_v$, then $\hat{f}(v^*) = f(g\rho_v^{-1}(v^*))$, and so $\hat{f}$ is regular on $\mathscr{U}_v$.

We have therefore shown that $\hat{f}$ is a regular function on $V_2^* \setminus \{0\}$, and therefore $\hat{f}$ is the restriction of a unique regular function $\tilde{f}$ on V_2^* which, by construction, verifies $\varphi(\tilde{f}) = f$. □

The proposition then follows immediately from Lemma 10.1.7 and the equalities $(*)$ and $(**)$. □

REMARK – The isomorphism $\mathbb{F}[\mathbf{G}]^{\rho(\mathbf{U})} \simeq \mathrm{Sym}(V_2)$ is in fact an algebraic consequence of the existence of a natural isomorphism of quasi-affine varieties $\mathbf{G}/\mathbf{U} \simeq V_2^* \setminus \{0\}$, $g\mathbf{U} \mapsto g \cdot y^*$. □

10.1.2. Simple Modules

We denote by $L(n)$ the $\mathbf{G}$-submodule of $\Delta(n)$ generated by x^n. Denote by

$$\begin{aligned} F_p \colon \quad \mathbf{G} &\longrightarrow \mathbf{G} \\ \begin{pmatrix} a & b \\ c & d \end{pmatrix} &\longmapsto \begin{pmatrix} a^p & b^p \\ c^p & d^p \end{pmatrix} \end{aligned}$$

the split Frobenius endomorphism of $\mathbf{G}$ over $\mathbb{F}_p$. If i is a non-negative integer and if V is a rational $\mathbf{G}$-module, we denote by $V^{(i)}$ the rational $\mathbf{G}$-module with the same underlying space, but on which the element $g \in \mathbf{G}$ acts as $F_p^i(g)$ acts on V. For example $V^{(0)} = V$. If $v \in V$, we will denote by $v^{(i)}$ the corresponding element of the $\mathbf{G}$-module $V^{(i)}$. If $g \in \mathbf{G}$, then

$$g \cdot v^{(i)} = (F_p^i(g) \cdot v)^{(i)}.$$

If m is a non-zero natural number, we will denote by $(m_i)_{i \geqslant 0}$ the unique sequence of elements of $\{0, 1, 2, \ldots, p-1\}$ such that

$$m = \sum_{i \geqslant 0} m_i p^i.$$

Note that the sequence $(m_i)_{i \geqslant 0}$ becomes zero after a certain point. Let

$$I(n) = \{m \in \mathbb{N} \mid \forall\, i \geqslant 0,\ m_i \leqslant n_i\}.$$

Note that $I(n) \subseteq \{0, 1, 2, \dots, n\}$ and that $0, n \in I(n)$. Moreover, $m \in I(n)$ if and only if $n - m \in I(n)$. The following theorem describes the simple rational **G**-modules as well as the simple $\mathbb{F}G$-modules.

Theorem 10.1.8. *With notation as above we have:*

(a) $L(n) = \oplus_{m \in I(n)} \mathbb{F} x^{n-m} y^m$. *In particular, if* $0 \leqslant n \leqslant p-1$, *then* $L(n) = \Delta(n)$.
(a′) *If* $0 \leqslant n \leqslant q-1$, *then* $L(n)$ *is the* G*-submodule of* $\mathrm{Res}^{\mathbf{G}}_{G} \Delta(n)$ *generated by* x^n.
(b) $L(n)$ *is the unique simple* **G***-submodule of* $\Delta(n)$.
(b′) *If* $0 \leqslant n \leqslant q-1$, *then* $\mathrm{Res}^{\mathbf{G}}_{G} L(n)$ *is the unique simple* $\mathbb{F}G$*-submodule of* $\mathrm{Res}^{\mathbf{G}}_{G} \Delta(n)$.
(c) $L(n) \simeq \bigotimes_{i \geqslant 0} L(n_i)^{(i)} = \bigotimes_{i \geqslant 0} \Delta(n_i)^{(i)}$.
(d) *If* $0 \leqslant m, n$, *then* $L(m) \simeq L(n)$ *if and only if* $m = n$.
(d′) *If* $0 \leqslant m, n \leqslant q-1$, *then* $\mathrm{Res}^{\mathbf{G}}_{G} L(m) \simeq \mathrm{Res}^{\mathbf{G}}_{G} L(n)$ *if and only if* $m = n$.
(e) *The family* $(L(n))_{n \geqslant 0}$ *is a set of representatives of the isomorphism classes of simple rational* **G***-modules.*
(e′) *The family* $(\mathrm{Res}^{\mathbf{G}}_{G} L(n))_{0 \leqslant n \leqslant q-1}$ *is a set of representatives of the isomorphism classes of simple* $\mathbb{F}G$*-modules.*

Proof. We begin by proving (a) and (a′). Let us temporarily denote by $V(n)$ the $\mathbb{F}$-subvector space of $\Delta(n)$ defined by

$$V(n) = \bigoplus_{m \in I(n)} \mathbb{F} x^{n-m} y^m.$$

Let $g = \begin{pmatrix} a & b \\ c & d \end{pmatrix} \in \mathbf{G}$ and let $m \in I(n)$. Then

$$g \cdot x^{n-m} y^m = (ax + cy)^{n-m} (bx + dy)^m = \prod_{i \geqslant 0} (ax + cy)^{p^i (n_i - m_i)} (bx + dy)^{p^i m_i}.$$

As a consequence,

$$(*) \qquad g \cdot x^{n-m} y^m = \prod_{i \geqslant 0} (a^{p^i} x^{p^i} + c^{p^i} y^{p^i})^{n_i - m_i} (b^{p^i} x^{p^i} + d^{p^i} y^{p^i})^{m_i}.$$

Expanding this product, we obtain a linear combination of monomials of the form

$$\prod_{i \geqslant 0} x^{p^i (n_i - m_i - s_i)} y^{p^i s_i} x^{p^i (m_i - t_i)} y^{p^i t_i},$$

where $0 \leqslant s_i \leqslant n_i - m_i$ and $0 \leqslant t_i \leqslant m_i$. Set $m' = \sum_{i \geqslant 0} (s_i + t_i) p^i$. Then $m' \in I(n)$ and

$$\prod_{i \geqslant 0} x^{p^i(n_i-m_i-s_i)} y^{p^i s_i} x^{p^i(m_i-t_i)} y^{p^i t_i} = x^{n-m'} y^{m'},$$

which shows that indeed $g \cdot x^{n-m} y^m \in V(n)$. Hence $V(n)$ is a **G**-submodule of $\Delta(n)$ containing x^n and therefore $V(n)$ contains $L(n)$.

To complete the proof of (a) and (a′), we may suppose that $0 \leqslant n \leqslant q-1$ (after possibly replacing q by a power). Now set

$$\begin{aligned} \mathbf{u}^-: \mathbb{F}_q^+ &\longrightarrow G \\ \xi &\longmapsto \begin{pmatrix} 1 & 0 \\ \xi & 1 \end{pmatrix} \end{aligned}$$

and denote by U^- the image of the morphism of groups $\mathbf{u}^-$. It will be sufficient for us to show that

(?) *If* $0 \leqslant n \leqslant q-1$, *then* $V(n)$ *is the* U^-*-submodule of* $\Delta(n)$ *generated by* x^n.

Now, if $\xi \in \mathbb{F}_q$, then it follows from $(*)$ that

$$\mathbf{u}^-(\xi) \cdot x^n = \sum_{m \in I(n)} \prod_{i \geqslant 0} \binom{n_i}{m_i} \xi^m x^{n-m} y^m.$$

If $m \in I(n)$, then $\prod_{i \geqslant 0} \binom{n_i}{m_i} \neq 0$, therefore it is sufficient for us to show that the matrix $(\xi^m)_{m \in I(n), \xi \in \mathbb{F}_q}$ is of rank $|I(n)|$, that is to say that the rows of this matrix are linearly independent. Now, as $0 \leqslant n \leqslant q-1$, these rows are complete rows of the matrix $(\xi^m)_{0 \leqslant m \leqslant q-1, \xi \in \mathbb{F}_q}$ which is a square Vandemonde matrix, and hence is clearly invertible.

The claims (b) and (b′) follow immediately from Corollary 10.1.5 and from (a′).

We now turn to (c). If $m \in I(n)$, we will denote by

$$v_m = \underset{i \geqslant 0}{\otimes} (x^{n_i - m_i} y^{m_i})^{(i)} \in \underset{i \geqslant 0}{\otimes} L(n_i)^{(i)} = \underset{i \geqslant 0}{\otimes} \Delta(n_i)^{(i)}.$$

The formula $(*)$ shows that the $\mathbb{F}$-linear map $L(n) \to \otimes_{i \geqslant 0} L(n_i)^{(i)}$ which sends $x^{n-m} y^m$ to v_m (for all $m \in I(n)$) and is in fact a morphism of **G**-modules. Moreover it is an isomorphism.

To show (d) and (d′) we may (after replacing q by a power), suppose that $0 \leqslant m \leqslant n \leqslant q-1$. Now, if $\operatorname{Res}_G^{\mathbf{G}} L(m) \simeq \operatorname{Res}_G^{\mathbf{G}} L(n)$, then the $\mathbb{F}T$-modules $L(m)^U$ and $L(n)^U$ are isomorphic. But T acts on $L(m)^U = \mathbb{F} x^m$ via the character $\operatorname{Res}_T^{\mathbf{T}} \varepsilon^m$. Hence, $\operatorname{Res}_T^{\mathbf{T}} \varepsilon^m = \operatorname{Res}_T^{\mathbf{T}} \varepsilon^n$ and therefore $m \equiv n \mod q-1$ because ε is injective on $T \simeq \mathbb{F}_q^\times$. As $0 \leqslant m, n \leqslant q-1$, this can only happen if $m = n$ or if $m = 0$ and $n = q-1$. Now $\dim_{\mathbb{F}} L(0) = 1 \neq q = \dim_{\mathbb{F}} L(q-1)$. Therefore $m = n$.

(e) We denote by $\mathbb{F}(n)$ the vector space $\mathbb{F}$ on which $\mathbf{B}$ acts via the linear character $\tilde{\varepsilon}^n$. Then $(\mathbb{F}(n))_{n\in\mathbb{Z}}$ is a family of representatives of the isomorphism classes of simple rational $\mathbf{B}$-modules. Indeed, if S is a simple rational $\mathbf{B}$-module, then $S^{\mathbf{U}} \neq 0$ (as $\mathbf{U}$ is a unipotent group [Bor, Theorems 4.4 and 4.8]) and is $\mathbf{B}$-stable, therefore $S = S^{\mathbf{U}}$ is a simple rational $\mathbf{T}$-module.

Let V be a simple rational $\mathbf{G}$-module. By the previous considerations there exists $n \in \mathbb{Z}$ and a surjective $\mathbf{B}$-equivariant morphism $\psi\colon \operatorname{Res}_{\mathbf{B}}^{\mathbf{G}} V \to \mathbb{F}(-n)$. Let

$$\begin{aligned} \tilde{\psi}\colon V &\longrightarrow \mathbb{F}[\mathbf{G}] \\ v &\longmapsto (g \mapsto \psi(g^{-1}\cdot v)). \end{aligned}$$

Then $\tilde{\psi}$ is a morphism of $\mathbf{G}$-modules (where $\mathbb{F}[\mathbf{G}]$ is equipped with the structure of a $\mathbf{G}$-module defined by $(g\cdot f)(x) = f(g^{-1}x)$ for all $g, x \in \mathbf{G}$ and $f \in \mathbb{F}[\mathbf{G}]$). As ψ is a morphism of $\mathbf{B}$-modules, the image of $\tilde{\psi}$ is contained in

$$\Delta'(n) = \{f \in \mathbb{F}[\mathbf{G}] \mid \forall\, (g,b) \in \mathbf{G}\times\mathbf{B},\ f(gb) = \tilde{\varepsilon}^{-n}(b)f(g)\}.$$

Moreover, $\tilde{\psi}\colon V \to \Delta'(n)$ is a non-zero morphism, and is therefore injective as V is simple. As $\Delta(n) \simeq \Delta'(n)$ (see Proposition 10.1.6), it follows from (b) that $V \simeq L(n)$.

The final statement (e′) now follows immediately from (d′) and from the fact that, by Proposition B.3.2(b) and Exercises 1.7 and 1.10, we have $|\operatorname{Irr} kG| = q$. □

COMMENTARY – In the case of a general connected reductive group, the classification of the simple rational modules by their highest weight is due to Borel and Weil. In the case of our group $\mathbf{G} = \mathrm{SL}_2(\mathbb{F})$, this corresponds to part (e) of Theorem 10.1.8. On the other hand, the description (in part (c) of Theorem 10.1.8) of the simple modules as tensor products of simple modules twisted by the action of the Frobenius also generalises to all connected reductive groups: this is the *Steinberg tensor product theorem*. For further details concerning the representations of reductive groups in positive characteristic the reader may consult the book of Jantzen [Jan].

A generalisation of part (e′) of Theorem 10.1.8 gives a parametrisation of the simple representations of finite reductive groups in equal characteristic: see, for example, [Cartier, §6]. □

EXAMPLE 10.1.9 – If $r \geqslant 0$, then $\Delta(p^r-1) = L(p^r-1)$. Indeed,

$$p^r - 1 = \sum_{i=0}^{r-1}(p-1)p^i$$

and therefore $I(p^r-1) = \{0,1,2,\dots,p^r-1\}$. □

10.1.3. The Grothendieck Ring of G

Theorem 10.1.8 shows that

(10.1.10)
$$\mathscr{K}_0(\mathbf{G}) = \bigoplus_{n \geqslant 0} \mathbb{Z}\,[L(n)]_{\mathbf{G}}.$$

On the other hand,

(10.1.11)
$$\mathrm{Car}_{\mathbf{G}}\, L(n) = \sum_{m \in I(n)} \varepsilon^{n-2m}.$$

Set

$$\varepsilon_n = \begin{cases} \varepsilon^0 & \text{if } n = 0, \\ \varepsilon^n + \varepsilon^{-n} & \text{if } n > 0. \end{cases}$$

Then

(10.1.12)
$$\mathrm{Car}_{\mathbf{G}}\, L(n) = \sum_{\substack{m \in I(n) \\ m \leqslant n/2}} \varepsilon_{n-2m} \in \varepsilon_n + \sum_{0 \leqslant m < n} \mathbb{Z}\varepsilon_m.$$

Now the element $s = \begin{pmatrix} 0 & -1 \\ 1 & 0 \end{pmatrix} \in \mathbf{G}$ normalises $\mathbf{T}$ and therefore acts on the Grothendieck group $\mathscr{K}_0(\mathbf{T})$. One may verify that

$$\mathscr{K}_0(\mathbf{T})^s = \bigoplus_{n \geqslant 0} \mathbb{Z}\varepsilon_n.$$

This shows that

(10.1.13) *The map* $\mathrm{Car}_{\mathbf{G}} : \mathscr{K}_0(\mathbf{G}) \to \mathscr{K}_0(\mathbf{T})$ *is injective with image* $\mathscr{K}_0(\mathbf{T})^s$.

Similarly, by 10.1.3, we have

(10.1.14)
$$\mathrm{Car}_{\mathbf{G}}\, \Delta(n) = \sum_{0 \leqslant m \leqslant n/2} \varepsilon_{n-2m} \in \varepsilon_n + \sum_{0 \leqslant m < n} \mathbb{Z}\varepsilon_m.$$

Which shows that $([\Delta(n)]_{\mathbf{G}})_{n \geqslant 0}$ is also a $\mathbb{Z}$-basis of the Grothendieck group $\mathscr{K}_0(\mathbf{G})$. If $m, n \geqslant 0$, we denote by $\boldsymbol{\Delta}_{m,n} = [\Delta(n) : L(m)]$ the multiplicity of $L(m)$ as a factor in a Jordan-Hölder series of $\Delta(n)$. These integers $\boldsymbol{\Delta}_{m,n}$ are uniquely determined by the equation

(10.1.15)
$$\mathrm{Car}_{\mathbf{G}}\, \Delta(n) = \sum_{m \geqslant 0} \boldsymbol{\Delta}_{m,n}\, \mathrm{Car}_{\mathbf{G}}\, L(m).$$

In particular, taking into account the formulas 10.1.12 and 10.1.14, we have

(10.1.16)
$$\boldsymbol{\Delta}_{m,n} \in \{0, 1\}$$

and

$$\mathbf{\Delta}_{m,n} = \begin{cases} 1 & \text{if } m = n, \\ 0 & \text{if } m > n, \\ 0 & \text{if } m \not\equiv n \mod 2. \end{cases} \tag{10.1.17}$$

Further to these equalities, it is not difficult to completely determine the infinite matrix $\mathbf{\Delta} = (\mathbf{\Delta}_{m,n})_{m,n \geqslant 0}$. For this we will need to define recursively a set $\mathscr{E}(n)$ as follows:

$$\mathscr{E}(n) = \begin{cases} \{0\} & \text{if } 0 \leqslant n \leqslant p-1, \\ p\mathscr{E}\left(\frac{n-n_0}{p}\right) & \text{if } n \geqslant p \text{ and } n_0 = p-1, \\ p\mathscr{E}\left(\frac{n-n_0}{p}\right) \dot{\bigcup} n_0+1+p\mathscr{E}\left(\frac{n-n_0-p}{p}\right) & \text{if } n \geqslant p \text{ and } n_0 \leqslant p-2. \end{cases}$$

The decomposition matrix is described in the next proposition.

Proposition 10.1.18. *Let m, $n \geqslant 0$. Then $\mathbf{\Delta}_{m,n} \in \{0,1\}$ and $\mathbf{\Delta}_{m,n} = 1$ if and only if $m \in n - 2\mathscr{E}(n)$.*

Proof. Let us first remark that:

If $r \in \mathscr{E}(n)$, then $n - 2r \geqslant 0$.

This follows from an elementary recurrence.

We will now show the proposition by induction on n. The crucial step is given by the following equality, in which we set $\Delta(-1) = 0$ to simplify notation. We claim that, if $n \geqslant 0$, then

$$\begin{aligned} [\Delta(n)]_{\mathbf{G}} = [\Delta(n_0)]_{\mathbf{G}} \cdot [\Delta\left(\tfrac{n-n_0}{p}\right)^{(1)}]_{\mathbf{G}} \\ +[\Delta(p-n_0-2)]_{\mathbf{G}} \cdot [\Delta\left(\tfrac{n-n_0-p}{p}\right)^{(1)}]_{\mathbf{G}}. \end{aligned} \tag{$*$}$$

We now show $(*)$. Write $n = n_0 + pn'$. We then have $(n-n_0)/p = n'$ and

$$\begin{aligned} \operatorname{Car}_{\mathbf{G}} \Delta(n_0) \otimes_{\mathbb{F}} \Delta(n')^{(1)} &= \sum_{i=0}^{n_0} \sum_{j=0}^{n'} \varepsilon^{n_0-2i} \varepsilon^{p(n'-2j)} \\ &= \sum_{i=0}^{n_0} \sum_{j=0}^{n'} \varepsilon^{n-2(i+pj)}. \end{aligned}$$

Moreover

$$\mathrm{Car}_{\mathbf{G}}\,\Delta(p-n_0-2)\otimes_{\mathbb{F}}\Delta(n'-1)^{(1)} = \sum_{i=0}^{p-n_0-2}\sum_{j=0}^{n'-1} \varepsilon^{p-n_0-2-2i}\varepsilon^{p(n'-1-2j)}$$
$$= \sum_{i=n_0+1}^{p-1}\sum_{j=0}^{n'-1} \varepsilon^{n-2(i+pj)}.$$

To show $(*)$, it is enough to remark that

$$\{0,1,2,\dots,n\} = \{i+pj \mid 0 \leqslant i \leqslant n_0 \text{ and } 0 \leqslant j \leqslant n'\}$$
$$\dot\cup\,\{i+pj \mid n_0+1 \leqslant i \leqslant p-1 \text{ and } 0 \leqslant j \leqslant n'-1\}.$$

We now return to the proof of Proposition 10.1.18. This result is immediate if $n \leqslant p-1$. Arguing by induction, we may suppose that $n \geqslant p$ and that the result holds for all $n'' < n$. By $(*)$ and the induction hypothesis, we have (as $L(i)\otimes L(j)^{(1)} = L(i+pj)$ if $0 \leqslant i \leqslant p-1$)

$$[\Delta(n)]_{\mathbf{G}} = [L(n_0)\otimes_{\mathbb{F}} \sum_{m\in\mathscr{E}(n')} L(n'-2m)^{(1)}]_{\mathbf{G}}$$
$$+\,[L(p-n_0-2)\otimes_{\mathbb{F}} \sum_{m\in\mathscr{E}(n'-1)} L(n'-1-2m)^{(1)}]_{\mathbf{G}}$$
$$= \begin{cases} \displaystyle\sum_{m\in\mathscr{E}(n')} [L(n-2pm)]_{\mathbf{G}} & \text{if } n_0 = p-1,\\[2ex] \displaystyle\sum_{m\in\mathscr{E}(n'-1)} [L(n-2(n_0+1+pm))]_{\mathbf{G}} & \text{if } n_0 \leqslant p-2.\end{cases}$$

The proof of the proposition is now complete. □

EXAMPLE 10.1.19 – If $i,j \in \{0,1,\dots,p-1\}$, then

$$[\Delta(pj+i)]_{\mathbf{G}} = \begin{cases} [L(pj+i)]_{\mathbf{G}} & \text{if } i = p-1 \text{ or } j = 0,\\ [L(pj+i)]_{\mathbf{G}} + [L(pj-i-2)]_{\mathbf{G}} & \text{if } i \leqslant p-2 \text{ and } j \geqslant 1.\end{cases}$$ □

10.2. Simple kG-Modules and Decomposition Matrices

This section is primarily concerned with determining the decomposition matrix of the finite group G modulo p.

10.2.1. *Simple kG-Modules*

Recall that $k \subseteq \mathbb{F}$ as k is of characteristic p. On the other hand, $\mathscr{O}$ contains the elements of the form $\xi+\xi^{-1}$, where ξ is a $(q+1)$-th root of unity (see the character table in Table 5.4). Therefore k contains all elements of the form $\xi+\xi^{-1}$, as ξ traverses μ_{q+1}. By virtue of Exercise 1.1, this implies that

$$\mathbb{F}_q \subseteq k \subseteq \mathbb{F}.$$

The $\mathbb{F}G$-module $\mathrm{Res}_G^{\mathbf{G}} \Delta(n)$ is the extension of scalars to $\mathbb{F}$ of a kG-module which we will denote by $\Delta_q(n)$: in fact, $\Delta_q(n) = \mathrm{Sym}^n(V_2^k)$, where $V_2^k = k^2$ is the natural representation of $G = \mathrm{SL}_2(\mathbb{F}_q) \subset \mathrm{GL}_2(k)$. Similarly, F_p restricts to an automorphism of the finite group G and we may define $\Delta_q(n)^{(i)}$ as previously, and similarly for $L_q(n)$ and $L_q(n)^{(i)}$. By Theorem 10.1.8(e′), the morphism

(10.2.1)
$$\begin{array}{ccc} \{0,1,2,\dots,q-1\} & \longrightarrow & \mathrm{Irr}\, kG \\ n & \longmapsto & [L_q(n)]_{kG} \end{array}$$

is bijective.

On the other hand, by 10.1.17, we have

(10.2.2)
$$[\Delta_q(n) : L_q(m)] = \mathbf{\Delta}_{m,n} = [\Delta(n) : L(m)]$$

for all $m, n \in \{0,1,2,\dots,q-1\}$, which shows that

(10.2.3)
$$\big([\Delta_q(n)]_{kG}\big)_{0 \leqslant n \leqslant q-1} \textit{ is a } \mathbb{Z}\textit{-basis of } \mathscr{K}_0(kG).$$

To finish this subsection, note the following result.

Lemma 10.2.4. *The simple kG-module $L_q(q-1)$ is projective. If M is an $\mathcal{O}G$-module such that $[KM]_{KG} = \mathrm{St}_G$, then $kM \simeq L_q(q-1)$.*

Proof. By the assertion (?) in the proof of Theorem 10.1.8,

$$L_q(q-1) = kU^- \cdot x^{q-1}.$$

As $\dim_k L_q(q-1) = q$ by 10.1.1 and Theorem 10.1.8(c), we deduce an isomorphism of kU^--modules $\mathrm{Res}_{U^-}^G L_q(q-1) \simeq kU^-$. Hence $\mathrm{Res}_{U^-}^G L_q(q-1)$ is a projective kU^--module, hence $L_q(q-1)$ is a projective kG-module (as U^- is a Sylow p-subgroup of G).

On the other hand, $L_q(q-1)^B = kx^{q-1} \neq 0$, and therefore there exists an injection $k \to \mathrm{Res}_B^G L_q(q-1)$ (where k denotes the trivial kB-module). Frobenius reciprocity then implies that there exists a non-zero morphism $\mathrm{Ind}_B^G k \to L_q(q-1)$. But $L_q(q-1)$ is simple and hence this morphism is surjective. The last assertion of the lemma follows. □

10.2.2. Decomposition Matrices

Recall that $\mathrm{Dec}(\mathcal{O}G)$ is the transpose of the matrix $\mathrm{dec}_{\mathcal{O}G} : \mathcal{K}_0(KG) \to \mathcal{K}_0(kG)$ in the bases $\mathrm{Irr}\, KG$ and $\mathrm{Irr}\, kG$. Taking into account 10.2.3, we denote by $\mathrm{Dec}_{\Delta}(\mathcal{O}G)$ the transpose of the matrix of the morphism $\mathrm{dec}_{\mathcal{O}G}$ in the bases $\mathrm{Irr}\, KG$ and $(\Delta_q(n))_{0 \leqslant n \leqslant q-1}$.

If we set $\mathbf{\Delta}[q]$ the matrix $(\mathbf{\Delta}_{m,n})_{0 \leqslant m,n \leqslant q-1}$ (that is to say, the matrix obtained from the infinite matrix $\mathbf{\Delta}$ by taking only the first q rows and columns), then

(10.2.5)
$$\mathrm{Dec}(\mathcal{O}G) = \mathrm{Dec}_{\Delta}(\mathcal{O}G)\, {}^t\mathbf{\Delta}[q].$$

As we have already determined the matrix $\mathbf{\Delta}[q]$ in Proposition 10.1.18 it is enough to determine the matrix $\mathrm{Dec}_{\Delta}(\mathcal{O}G)$.

To this end, note that the canonical surjection $\mathcal{O}^\times \to k^\times$ admits a unique section which we denote by $\kappa : k^\times \hookrightarrow \mathcal{O}^\times$ (see 6.1.1). For $H \in \{T, T'\}$ let

$$\begin{aligned} \varepsilon_H :\ H &\longrightarrow \mathcal{O}^\times \\ t &\longmapsto \kappa(\varepsilon(t)). \end{aligned}$$

We identify $\mathcal{K}_0(kG)$ with the lattice of (virtual) Brauer characters [CuRe, Definition 17.4 and Proposition 17.14]. Now a Brauer character β of G (which is defined on the p-regular elements of G, that is to say the set of semi-simple elements) is totally determined by the restrictions $\mathrm{Res}_T^G \beta$ and $\mathrm{Res}_{T'}^G \beta$.

For example, it follows from 10.1.3 that

(10.2.6)
$$\mathrm{Res}_H^G [\Delta_q(n)]_{kG} = \sum_{i=0}^{n} \varepsilon_H^{n-2i}$$

for all $0 \leqslant n \leqslant q-1$.

On the other hand, if $H \in \{T, T'\}$, then

$$\mathrm{Irr}\, H = \{\varepsilon_H^i \mid 0 \leqslant i \leqslant |H|-1\}.$$

Examining the character table of G, we may easily verify that, if $j \in \mathbb{Z}$, then

(10.2.7)
$$\mathrm{Res}_H^G R(\varepsilon_T^j) = \delta_{H=T}(\varepsilon_H^j + \varepsilon_H^{-j}) + 2 \sum_{\substack{0 \leqslant i \leqslant |H|-1 \\ i \equiv j \mod 2}} \varepsilon_H^i$$

and

(10.2.8)
$$\mathrm{Res}_H^G R'(\varepsilon_{T'}^j) = -\delta_{H=T'}(\varepsilon_H^j + \varepsilon_H^{-j}) + 2 \sum_{\substack{0 \leqslant i \leqslant |H|-1 \\ i \equiv j \mod 2}} \varepsilon_H^i.$$

Here, $\delta_{H=T}$ is equal to 0 if $H = T'$ and 1 if $H = T$ (and, similarly, $\delta_{H=T'} = 1 - \delta_{H=T}$). We may easily deduce (again using the character table of G) the form of the matrix $\mathrm{Dec}_\Delta(\mathcal{O}G)$:

Proposition 10.2.9. *To simply notation set $\Delta_q(-1) = L_q(-1) = 0$. Then:*
(a) $\mathrm{dec}_{\mathcal{O}G} 1_G = [\Delta_q(0)]_{kG} = [L_q(0)]_{kG}$.
(b) $\mathrm{dec}_{\mathcal{O}G} \mathrm{St}_G = [\Delta_q(q-1)]_{kG} = [L_q(q-1)]_{kG}$.
(c) *If $0 \leqslant i \leqslant q-1$, then* $\mathrm{dec}_{\mathcal{O}G} \mathrm{R}(\varepsilon_T^i) = [\Delta_q(i)]_{kG} + [\Delta_q(q-1-i)]_{kG}$.
(d) *If $1 \leqslant i \leqslant q$, then* $\mathrm{dec}_{\mathcal{O}G} \mathrm{R}'(\varepsilon_{T'}^i) = [\Delta_q(i-2)]_{kG} + [\Delta_q(q-1-i)]_{kG}$.
(e) $\mathrm{dec}_{\mathcal{O}G} \mathrm{R}_\pm(\varepsilon_T^{(q-1)/2}) = [\Delta_q((q-1)/2)]_{kG}$.
(f) $\mathrm{dec}_{\mathcal{O}G} \mathrm{R}'_\pm(\varepsilon_{T'}^{(q+1)/2}) = [\Delta_q((q-3)/2)]_{kG}$.

10.3. Blocks

10.3.1. Blocks and Brauer Correspondents

Recall that $\mathbf{d}$ denotes the natural isomorphism of groups $\mathbb{F}_q^\times \xrightarrow{\sim} T$. We set

$$e_+ = \frac{1}{2}(\mathbf{d}(1) + \mathbf{d}(-1)), \quad e_+^G = e_+ - e_{\mathrm{St}_G} \quad \text{and} \quad e_-^G = e_- = \frac{1}{2}(\mathbf{d}(1) - \mathbf{d}(-1)).$$

Then it is straightforward to verify that

$$1 = e_+^G + e_-^G + e_{\mathrm{St}_G} \tag{10.3.1}$$

is a decomposition of 1 as a sum of pairwise orthogonal central idempotents of $\mathcal{O}G$. The following proposition is even better:

Proposition 10.3.2. *The central idempotents e_+^G, e_-^G and e_{St_G} are primitive.*

Proof. Denote by 1_+ (respectively 1_-) the trivial (respectively non-trivial) linear character of Z. If $\chi \in \mathrm{Irr}\, G$, we set $\tilde{\omega}_\chi$ the unique linear character of Z such that $\chi(zg) = \tilde{\omega}_\chi(z)\chi(g)$ for all $(g, z) \in G \times Z$. Let

$$\mathscr{B}_+ = \{\chi \in \mathrm{Irr}\, G \mid \tilde{\omega}_\chi = 1_+\} \setminus \{\mathrm{St}_G\},$$

$$\mathscr{B}_- = \{\chi \in \mathrm{Irr}\, G \mid \tilde{\omega}_\chi = 1_-\}$$

and

$$\mathscr{B}_{\mathrm{St}} = \{\mathrm{St}_G\}.$$

One may easily verify, thanks to the table of central characters in Table 6.1, that $\mathscr{B}_+$, $\mathscr{B}_-$ and $\mathscr{B}_{\mathrm{St}}$ are the p-blocks of G, which shows the result. □

We set

$$A_+ = \mathscr{O}Ge_+^G, \quad A_- = \mathscr{O}Ge_-^G \quad \text{and} \quad A_{\mathrm{St}} = \mathscr{O}Ge_{\mathrm{St}_G}.$$

Note that A_+ is the principal $\mathscr{O}$-block of **G** and that A_{St} is a block with trivial defect. The straightforward calculation of the Brauer correspondence is left to the reader.

Proposition 10.3.3. *The group U is a defect group of A_+ and A_-, and $\mathscr{O}Be_?$ is a Brauer correspondent of $A_?$ (for $? \in \{+,-\}$).*

Proposition 10.3.4. *We have*

$$\operatorname{Irr} kA_+ = \{[L_q(2n)]_{kG} \mid 0 \leqslant n \leqslant (q-3)/2\},$$

$$\operatorname{Irr} kA_- = \{[L_q(2n+1)]_{kG} \mid 0 \leqslant n \leqslant (q-3)/2\}$$

and

$$\operatorname{Irr} kA_{\mathrm{St}} = \{[L_q(q-1)]_{kG}\}.$$

Broué's conjecture was shown in this case by Okuyama [Oku1], [Oku2] for the principal block and by Yoshii [Yo] for the nonpricipal one.

Theorem 10.3.5 (Okuyama, Yoshii). *There exist Rickard equivalences*

$$\mathrm{D}^b(A_+) \simeq \mathrm{D}^b(\mathscr{O}Be_+) \quad \textit{and} \quad \mathrm{D}^b(A_-) \simeq \mathrm{D}^b(\mathscr{O}Be_-).$$

10.3.2. Brauer Trees*

As Brauer trees are only defined in the case of cyclic defect group we make the following assumption which guarantees that the group U, which is the defect group of A_+ and A_-, is cyclic.

Hypothesis. *In this and only this subsection we suppose that $\ell = p = q$.*

It follows from this hypothesis that

$$\boldsymbol{\Delta}[q] = \boldsymbol{\Delta}[p] = \mathrm{I}_p,$$

and therefore that

$$\mathrm{Dec}(\mathscr{O}G) = \mathrm{Dec}_{\Delta}(\mathscr{O}G)$$

(see 10.2.5). Let $R_0 = 1_G$, $R_i = \mathrm{R}(\varepsilon_T^i)$ (if $1 \leqslant i \leqslant (p-1)/2$) and $R'_i = \mathrm{R}'(\varepsilon_{T'}^i)$ (if $1 \leqslant i \leqslant (p+1)/2$). Up to the planar embedding, the Brauer tree of a block with cyclic defect group is completely determined by its decomposition matrix. We conclude from Proposition 10.2.9 that the Brauer trees of A_+ and A_- are as follows.

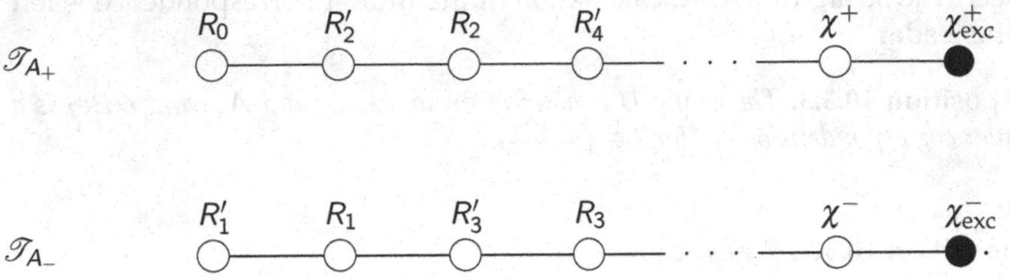

where

$$\chi^+ = \begin{cases} R'_{(p-1)/2} & \text{if } p \equiv 1 \mod 4, \\ R_{(p-3)/2} & \text{if } p \equiv 3 \mod 4, \end{cases} \quad \text{and } \chi^+_{\mathrm{exc}} = \begin{cases} R_{(p-1)/2} & \text{if } p \equiv 1 \mod 4, \\ R'_{(p+1)/2} & \text{if } p \equiv 3 \mod 4, \end{cases}$$

and, similarly,

$$\chi^- = \begin{cases} R_{(p-3)/2} & \text{if } p \equiv 1 \mod 4, \\ R'_{(p-1)/2} & \text{if } p \equiv 3 \mod 4, \end{cases} \quad \text{and } \chi^-_{\mathrm{exc}} = \begin{cases} R'_{(p+1)/2} & \text{if } p \equiv 1 \mod 4, \\ R_{(p-1)/2} & \text{if } p \equiv 3 \mod 4. \end{cases}$$

We can also calculate the Brauer trees for the Brauer correspondents $A_+^\flat$ and $A_-^\flat$ of A_+ and A_- respectively. The exceptional characters of $A_+^\flat$ (respectively $A_-^\flat$) are $\chi^B_{+,+}$ and $\chi^B_{+,-}$ (respectively $\chi^B_{-,+}$ and $\chi^B_{-,-}$). Using the character table of the group B, we obtain the following Brauer trees.

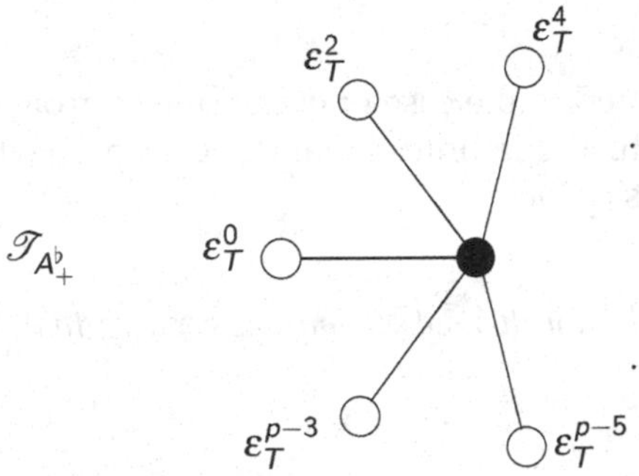

$\mathcal{T}_{A^\flat_-}$

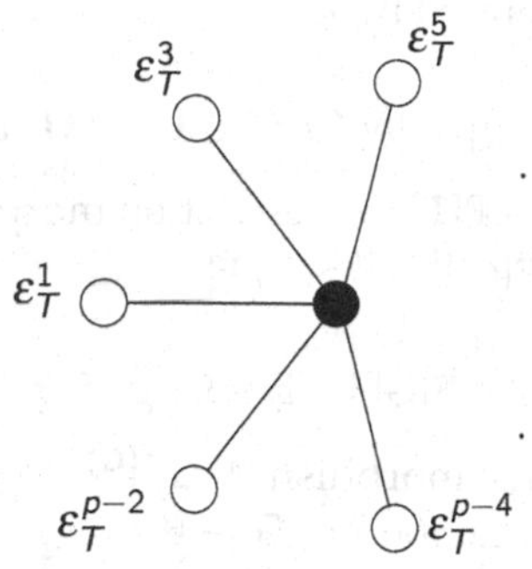

By virtue of Brauer's theorem (see Appendix C), we may conclude that, if $? \in \{+,-\}$, then

(10.3.6) *the blocks $A_?$ and $A_?^\flat$ are not Morita equivalent.*

REMARK – The blocks A_+ and A_- are Morita equivalent because they have the same Brauer tree (Brauer's theorem) while the $\mathcal{O}$-algebras $A_+^\flat$ and $A_-^\flat$ are isomorphic (by restriction of the isomorphism $\mathcal{O}B \to \mathcal{O}B$, $b \mapsto \varepsilon_T(b)b$). □

Exercises

10.1* (Carter-Lusztig). Denote by $\mathcal{K}_1(kG)$ the Grothendieck group of the category of projective kG-modules, which we may view as a sublattice (and even an ideal) of $\mathcal{K}_0(KG)$. Show that

$$\mathcal{K}_1(kG) = \mathcal{K}_0(KG) \cdot \mathrm{St}_G\,.$$

10.2. If $\alpha\colon T \to k^\times$ is a linear character, we denote by k_α the kT-module defined as follows: the underlying k-vector space is k itself and an element $t \in T$ acts as multiplication by $\alpha(t)$. Show that, if $\alpha \neq \alpha^{-1}$, then the kG-modules $k[G/U] \otimes_{kT} k_\alpha$ and $k[G/U] \otimes_{kT} k_{\alpha^{-1}}$ are not isomorphic.

Compare with Exercise 7.4.

10.3*. The purpose of this exercise is to calculate certain subalgebras of invariants.

(a) Let

$$\mathbb{F}[\mathbf{G}]^{\mathbf{U}\times\mathbf{U}} = \{f \in \mathbb{F}[\mathbf{G}] \mid \forall\,(u,v) \in \mathbf{U}\times\mathbf{U},\ u\cdot f\cdot v = f\}$$

and denote by $\varphi\colon \mathbf{G} \to \mathbb{F}$ the regular function defined by

$$\varphi\begin{pmatrix} a & b \\ c & d \end{pmatrix} = c.$$

Show that $\mathbb{F}[\mathbf{G}]^{\mathbf{U}\times\mathbf{U}}$ is generated by φ.

(b) Let

$$\mathbb{F}[\mathbf{G}]^{\mathbf{U}^-\times\mathbf{U}} = \{f \in \mathbb{F}[\mathbf{G}] \mid \forall\,(u,v) \in \mathbf{U}^- \times \mathbf{U},\ u\cdot f\cdot v = f\}$$

and denote by $\mathrm{res}_{\mathbf{T}} : \mathbb{F}[\mathbf{G}] \to \mathbb{F}[\mathbf{T}]$ the restriction morphism. Show that $\mathrm{res}_{\mathbf{T}}$ induces an isomorphism $\mathbb{F}[\mathbf{G}]^{\mathbf{U}^-\times\mathbf{U}} \simeq \mathbb{F}[\mathbf{T}]$.

(c) Let

$$\mathbb{F}[\mathbf{G}]^{\gamma(\mathbf{G})} = \{f \in \mathbb{F}[\mathbf{G}] \mid \forall\, g \in \mathbf{G},\ g\cdot f\cdot g^{-1} = f\}.$$

Show that $\mathrm{res}_{\mathbf{T}}$ induces an isomorphism $\mathbb{F}[\mathbf{G}]^{\gamma(\mathbf{G})} \simeq \mathbb{F}[\mathbf{T}]^s$ and that $\mathbb{F}[\mathbf{G}]^{\gamma(\mathbf{G})}$ is generated by the trace function $\mathrm{Tr}\colon \mathbf{G} \to \mathbb{F}$.

(d) Let

$$\mathbb{F}[\mathbf{G}]^{\rho(\mathbf{T})} = \{f \in \mathbb{F}[\mathbf{G}] \mid \forall\, t \in \mathbf{T},\ f\cdot t = f\}.$$

If $g = \begin{pmatrix} a & b \\ c & d \end{pmatrix} \in \mathbf{G}$, we set

$$f_1(g) = ab, \quad f_2(g) = cd, \quad d_1(g) = ad \quad \text{and} \quad d_2(g) = bc.$$

Show that $\mathbb{F}[\mathbf{G}]^{\rho(\mathbf{T})}$ is generated by f_1,f_2, d_1 and d_2 and that the kernel of the canonical morphism

$$\begin{array}{ccc} \mathbb{F}[F_1,F_2,D_1,D_2] & \longrightarrow & \mathbb{F}[\mathbf{G}]^{\rho(\mathbf{T})} \\ P & \longmapsto & P(f_1,f_2,d_1,d_2) \end{array}$$

is generated by $F_1F_2 - D_1D_2$ and $D_1 - D_2 - 1$ (here, F_1, F_2, D_1 and D_2 are indeterminates).

Part IV
Complements

The goals of these two final chapters are totally opposite. In Chapter 11, we consider the results obtained for $\mathsf{SL}_2(\mathbb{F}_q)$ when $q \in \{3, 5, 7\}$ and point out various curious facts that occur in these small cases. We will study morphisms to $\mathsf{SL}_2(\mathbb{F}_r)$ as well as encountering exceptional reflection groups of ranks 2 and 3, the Klein curve and certain exceptional isomorphisms. In Chapter 12, we give a very brief summary (without proof) of Deligne-Lusztig theory, whose goal is the study of representations of arbitrary finite reductive groups using geometric methods. We explain to what extent the results that we have seen for $\mathsf{SL}_2(\mathbb{F}_q)$ may be interpreted as a special case of the general theory.

Chapter 11
Special Cases

In this chapter we will make explicit certain exotic properties of the groups $\mathsf{SL}_2(\mathbb{F}_q)$ when $q = 3$, 5 or 7. These include exceptional isomorphisms, inclusions as subgroups of $\mathsf{SL}_2(\mathbb{F}_q)$, and realisations as subgroups of reflection groups. For a recollection of definitions, results about reflection groups, see the Appendix C.

Notation. *We will denote by Z the centre of $G = \mathsf{SL}_2(\mathbb{F}_q)$, $\widetilde{Z}$ the centre of $\widetilde{G} = \mathsf{GL}_2(\mathbb{F}_q)$, $\mathsf{PGL}_2(\mathbb{F}_q) = \widetilde{G}/\widetilde{Z}$ the projective general linear group and $\mathsf{PSL}_2(\mathbb{F}_q) = G/Z$ the projective special linear group. If n is a non-zero natural number, we denote by $\mathfrak{S}_n$ the symmetric group of degree n (i.e. the group of permutations of the set $\{1,2,\ldots,n\}$) and $\varepsilon_n \colon \mathfrak{S}_n \to \{1,-1\}$ the sign homomorphism. The alternating group of degree n (the kernel of the sign homomorphism) will be denoted $\mathfrak{A}_n$.*

11.1. Preliminaries

The linear character α_0 of $\mathbb{F}_q^\times$, of order 2, is the *Legendre character*: $\alpha_0(z)$ is equal to 1 if z is a square and is -1 otherwise. In particular, the linear character $\alpha_0 \circ \det$ of $\widetilde{G}$ factorises to give a character of $\widetilde{G}/\widetilde{Z}$ which we will denote by $\det_0 \colon \mathsf{PGL}_2(\mathbb{F}_q) \to \{1,-1\}$. One may easily verify that

(11.1.1) $$\mathsf{PSL}_2(\mathbb{F}_q) = \mathrm{Ker}(\det_0).$$

The action of $\mathsf{PGL}_2(\mathbb{F}_q)$ on the finite projective line $\mathbf{P}^1(\mathbb{F}_q)$ induces, after enumerating its elements 1 up to $q+1$, a homomorphism of groups

C. Bonnafé, *Representations of* $\mathsf{SL}_2(\mathbb{F}_q)$, Algebra and Applications 13,
DOI 10.1007/978-0-85729-157-8_11, © Springer-Verlag London Limited 2011

$$\phi_q\colon \mathsf{PGL}_2(\mathbb{F}_q) \longrightarrow \mathfrak{S}_{q+1}.$$

The character $\det_0$ may be expressed with the help of the sign homomorphism on $\mathfrak{S}_{q+1}$:

(11.1.2) $$\det_0 = \varepsilon_{q+1} \circ \phi_q.$$

Proof. Recall that $\widetilde{B}$ denotes the subgroup of $\widetilde{G}$ formed by upper triangular matrices. As $\widetilde{G} = \widetilde{B} \cup \widetilde{B}s\widetilde{B}$, the group $\widetilde{G}$ is generated by the conjugates of $\widetilde{B}$. As a consequence, it is enough to show that, if $g \in \widetilde{B}$, then

$$\alpha_0(\det(g)) = \varepsilon_{q+1} \circ \phi_q(\bar{g}),$$

where $\bar{g}$ denotes the image of g in $\mathsf{PGL}_2(\mathbb{F}_q)$. Write

$$g = \begin{pmatrix} a & b \\ 0 & c \end{pmatrix}.$$

Let $\infty = [1;0] \in \mathbf{P}^1(\mathbb{F}_q)$. Then $g \cdot \infty = \infty$ and, if $z \in \mathbb{F}_q$, then $g \cdot [z;1] = [(az+b)/c;1]$. Denote by $\sigma\colon \mathbb{F}_q \to \mathbb{F}_q$, $z \mapsto (az+b)/c$. It suffices to show that the sign of the permutation σ is equal to $\alpha_0(ac)$.

To this end, fix a total order $\leqslant$ on $\mathbb{F}_q$ and set

$$A = \prod_{\substack{z,z' \in \mathbb{F}_q \\ z<z'}} (z'-z) \in \mathbb{F}_q^\times.$$

Then the sign ε of σ is defined by

$$\varepsilon A = \prod_{\substack{z,z' \in \mathbb{F}_q \\ z<z'}} (\sigma(z') - \sigma(z)).$$

It follows that, in $\mathbb{F}_q^\times$, we have $\varepsilon = (a/c)^{q(q-1)/2} = (ac)^{q(q-1)/2} c^{-q(q-1)} = (ac)^{(q-1)/2}$. Hence $\varepsilon = \alpha_0(ac)$ as expected. $\square$

Hence the morphism ϕ_q induces by restriction a morphism

$$\phi'_q\colon \mathsf{PSL}_2(\mathbb{F}_q) \longrightarrow \mathfrak{A}_{q+1}.$$

Recall that

(11.1.3) *ϕ_q and ϕ'_q are injective*

and

(11.1.4) $$|\mathsf{PGL}_2(\mathbb{F}_q)| = q(q^2-1) \quad\text{and}\quad |\mathsf{PSL}_2(\mathbb{F}_q)| = \frac{q(q^2-1)}{2}.$$

11.2. The Case when $q = 3$

Hypothesis. *In this and only this section we suppose that $q = 3$.*

11.2.1. Structure

Recall that

(11.2.1) $$|\mathrm{SL}_2(\mathbb{F}_3)| = 24, \quad |\mathrm{PGL}_2(\mathbb{F}_3)| = 24 \quad \text{and} \quad |\mathrm{PSL}_2(\mathbb{F}_3)| = 12,$$

which shows that the morphisms ϕ_3 and ϕ'_3 defined in the previous sections are isomorphisms, that is to say

(11.2.2) $$\mathrm{PGL}_2(\mathbb{F}_3) \simeq \mathfrak{S}_4 \quad \text{and} \quad \mathrm{PSL}_2(\mathbb{F}_3) \simeq \mathfrak{A}_4.$$

Therefore $\mathrm{SL}_2(\mathbb{F}_3)$ is a non-trivial central extension of $\mathfrak{A}_4$ by $\mathbb{Z}/2\mathbb{Z}$ (which is unique up to isomorphism) while $\mathrm{PGL}_2(\mathbb{F}_3)$ is a non-trivial central extension of $\mathfrak{S}_4$ (there are three such extensions as $H^2(\mathfrak{S}_4, \mathbb{Z}/2\mathbb{Z}) \simeq \mathbb{Z}/2\mathbb{Z} \times \mathbb{Z}/2\mathbb{Z}$).

We now show the following result, stated in Theorem 1.2.4.

Proposition 11.2.3. *We have $G = N' \rtimes U$ and $N' = \mathrm{D}(G)$. The group N' is isomorphic to the quaternionic group of order* 8. *The only non-trivial normal subgroups of G are Z and N'.*

Proof. We have $|N'| = 8$ and $|G| = 24$, and therefore N' is a Sylow 2-subgroup of G. It is therefore the inverse image, under ϕ'_3, of the Sylow 2-subgroup of $\mathfrak{A}_4$. Now the Sylow 2-subgroup of $\mathfrak{A}_4$ is normal in $\mathfrak{A}_4$ (and even in $\mathfrak{S}_4$), therefore N' is normal in G. On the other hand, as $N' \cap U = \{\mathrm{I}_2\}$ and $|N'| \cdot |U| = |G|$, we have $G = N' \rtimes U$.

This shows in particular that $\mathrm{D}(G) \subseteq N'$. The equality $N' = \mathrm{D}(G)$ and the two last assertions of the proposition are left as an (easy) exercise. □

Let

$$I = \begin{pmatrix} 0 & -1 \\ 1 & 0 \end{pmatrix}, \quad J = \begin{pmatrix} 1 & 1 \\ 1 & -1 \end{pmatrix} \quad \text{and} \quad K = \begin{pmatrix} -1 & 1 \\ 1 & 1 \end{pmatrix}.$$

Then

$$N' = \{\pm \mathrm{I}_2, \pm I, \pm J, \pm K\}$$

and the multiplication is given by

$$I^2 = J^2 = K^2 = -\mathrm{I}_2, \quad IJ = -JI = K, \quad JK = -KJ = I \quad \text{and} \quad KI = -IK = J,$$

which shows again that N' is the quaternionic group of order 8. To simplify notation let

$$u = u_+ = \begin{pmatrix} 1 & 1 \\ 0 & 1 \end{pmatrix}.$$

One may easily verify that

$$U = \langle u \rangle, \quad u^3 = \mathrm{I}_2, \quad {}^u I = J, \quad {}^u J = K \quad \text{and} \quad {}^u K = I.$$

11.2.2. Character Table

The character table 5.4 specialises as follows. Let i denote an element of μ_4 of order 4 (and note that $q_0 = -3$), let $j = (-1+\sqrt{-3})/2$ (so that j is a third root of unity in K) and let $i^\wedge$ be a linear character of order 4 of μ_4. The character table of $\mathrm{SL}_2(\mathbb{F}_3)$ is given in Table 11.1.

Table 11.1 Character table of $\mathrm{SL}_2(\mathbb{F}_3)$

g	I_2	$-\mathrm{I}_2$	$\mathbf{d}'(i)$	u_+	u_-	$-u_+$	$-u_-$
$\lvert \mathrm{Cl}_G(g) \rvert$	1	1	6	4	4	4	4
$o(g)$	1	2	4	3	3	6	6
$\mathrm{C}_G(g)$	G	G	T'	ZU	ZU	ZU	ZU
1_G	1	1	1	1	1	1	1
$\mathrm{R}'_+(\theta_0)$	1	1	1	j	j^2	j	j^2
$\mathrm{R}'_-(\theta_0)$	1	1	1	j^2	j	j^2	j
$\mathrm{R}'(i^\wedge)$	2	-2	0	-1	-1	1	1
$\mathrm{R}_+(\alpha_0)$	2	-2	0	$-j^2$	$-j$	j^2	j
$\mathrm{R}_-(\alpha_0)$	2	-2	0	$-j$	$-j^2$	j	j^2
St_G	3	3	-1	0	0	0	0

11.2.3. The Group $\mathrm{SL}_2(\mathbb{F}_3)$ as a Subgroup of $\mathrm{SL}_2(\mathbb{F}_\ell)$

Amongst the three characters of degree 2 of G, only $\mathrm{R}'(i^\wedge)$ has rational values. Here we study the possible fields of definition of a representation admitting this character. For this, we commence by calculating the Frobenius-Schur indicator of $\mathrm{R}'(i^\wedge)$:

$$\frac{1}{|G|}\sum_{g\in G}\mathrm{R}'(i^\wedge)(g^2)$$
$$=\frac{1}{24}(2+2+6\times(-2)+4\times(-1)+4\times(-1)+4\times(-1)+4\times(-1))=-1.$$

As a consequence (see [Isa, Corollary 4.15]), $\mathrm{R}'(i^\wedge)$ is not the character of a *real* representation, even though it has rational (and hence real) values.

Proposition 11.2.4. *Let R be a ring in which 2 is invertible and -1 is a sum of two squares. Then there exists an injective homomorphism $\rho : \mathrm{SL}_2(\mathbb{F}_3) \to \mathrm{SL}_2(R)$ sending $-\mathrm{I}_2$ to $\begin{pmatrix} -1 & 0 \\ 0 & -1 \end{pmatrix}$ and such that, if $g \in \mathrm{SL}_2(\mathbb{F}_3)$, then*

$$\mathrm{Tr}(\rho(g)) = \mathrm{R}'(i^\wedge)(g)\cdot 1_R \in R.$$

Proof. Choose a and b in R such that $a^2+b^2=-1$. Write

$$\rho(I)=\begin{pmatrix} 0 & -1 \\ 1 & 0 \end{pmatrix},\quad \rho(J)=\begin{pmatrix} a & b \\ b & -a \end{pmatrix},\quad \rho(K)=\begin{pmatrix} -b & a \\ a & b \end{pmatrix}$$

and

$$\rho(u)=\frac{1}{2}\begin{pmatrix} b-a-1 & 1-a-b \\ -1-a-b & a-b-1 \end{pmatrix}.$$

One verifies easily that $\det\rho(u)=1$, $\mathrm{Tr}\,\rho(u)=-1$ (and therefore $\rho(u)^2+\rho(u)+1=0$), that

$$\rho(I)^2=\rho(J)^2=\rho(K)^2=\begin{pmatrix} -1 & 0 \\ 0 & -1 \end{pmatrix},$$

that

$$\rho(I)\rho(J)=-\rho(J)\rho(I)=\rho(K),$$

and that

$$^{\rho(u)}\rho(I)=\rho(J),\quad {}^{\rho(u)}\rho(I)=\rho(J)\quad\text{and}\quad {}^{\rho(u)}\rho(I)=\rho(J).$$

It follows that we can extend ρ to a morphism of groups

$$\rho:\ \mathrm{SL}_2(\mathbb{F}_3)\longrightarrow \mathrm{SL}_2(R)$$

satisfying the conditions of the proposition. □

Corollary 11.2.5. *Let R be a principal ideal domain of characteristic different from 2 and let $\mathbb{K}$ be its field of fractions. We assume that -1 is the sum of two squares in $\mathbb{K}$. Then there exists an injective morphism of groups $\rho\colon \mathsf{SL}_2(\mathbb{F}_3) \to \mathsf{SL}_2(R)$ sending $-\mathsf{I}_2$ to $\begin{pmatrix} -1 & 0 \\ 0 & -1 \end{pmatrix}$ and such that, if $g \in \mathsf{SL}_2(\mathbb{F}_3)$, then*

$$\mathsf{Tr}(\rho(g)) = \mathsf{R}'(i^{\wedge})(g) \cdot 1_R \in R.$$

Proof. By Proposition 11.2.4, there exists an injective morphism of groups $\rho\colon \mathsf{SL}_2(\mathbb{F}_3) \to \mathsf{SL}_2(\mathbb{K})$ sending $-\mathsf{I}_2$ to $\begin{pmatrix} -1 & 0 \\ 0 & -1 \end{pmatrix}$. The proposition follows from Proposition B.3.1 (as R is a principal ideal domain). □

The corollary shows that, if $\mathbb{K}$ is a field of characteristic different from 2 in which -1 is a sum of two squares, then there exists a faithful irreducible representation $\mathsf{SL}_2(\mathbb{F}_3)$ over $\mathbb{K}$ of dimension 2 admitting $\mathsf{R}'(i^{\wedge})$ as character. In fact the converse is also true (see Exercise 11.4).

EXAMPLE 11.2.6 – Denote by ζ_n a primitive n^{th} root of unity in $\mathbb{C}$. Then $\mathbb{Z}[\zeta_3], \mathbb{Z}[\zeta_4]$ and $\mathbb{Z}[\zeta_5]$ satisfy the hypotheses of Corollary 11.2.5. Indeed, $-1 = \zeta_3^2 + (\zeta_3^2)^2 = \zeta_4^2 + 0^2 = (\zeta_5 + \zeta_5^{-1})^2 + (\zeta_5^2 - \zeta_5^{-2})^2$. □

EXAMPLE 11.2.7 – In a finite field every element is the sum of two squares. This is immediate in characteristic 2. If $\mathbb{K}$ is a finite field of characteristic different from 2 and cardinality r, then, if we let $C = \{a^2 \mid a \in \mathbb{K}\}$, we have $|C| = (r+1)/2$. Therefore, if $z \in \mathbb{K}$, the sets C and $z - C$ cannot be disjoint. Therefore $C + C = \mathbb{K}$ and, in particular, -1 is the sum of two squares.

As a consequence, for all odd prime numbers ℓ, -1 is the sum of two squares in $\mathbb{F}_\ell$ and, by Hensel's lemma, -1 is the sum of two squares in $\mathbb{Z}_\ell$. We then obtain two injective group homomorphisms $\mathsf{SL}_2(\mathbb{F}_3) \hookrightarrow \mathsf{SL}_2(\mathbb{Z}_\ell)$ and $\rho_{3\to\ell}\colon \mathsf{SL}_2(\mathbb{F}_3) \hookrightarrow \mathsf{SL}_2(\mathbb{F}_\ell)$. We may then deduce a new proof of Theorem 1.4.3(d): if $\ell \equiv \pm 3 \mod 8$, then $S = \rho_{3\to\ell}(N')$ is a Sylow 2-subgroup of $\mathsf{SL}_2(\mathbb{F}_\ell)$. Therefore, the image of $\rho_{3\to\ell}$ is contained in $\mathsf{N}_{\mathsf{SL}_2(\mathbb{F}_\ell)}(S)$, which shows that $|\mathsf{N}_{\mathsf{SL}_2(\mathbb{F}_\ell)}(S)/S| \geqslant 3$. □

11.2.4. *The Group* $\mathsf{SL}_2(\mathbb{F}_3)$ *as a Reflection Group of Rank* 2

A review of reflection groups is contained in Appendix C. Let V be a $\mathbb{C}$-vector space of dimension 2 and let $\rho\colon G \to \mathsf{GL}_\mathbb{C}(V)$ be an irreducible representation of G whose character does not take on only rational values. If u is an element of G of order 3, and if we denote by ζ, ζ' the eigenvalues of $\rho(u)$, then ζ and ζ' are third roots of unity such that $\zeta + \zeta' \in \{-j, -j^2\}$.

Consequently, $\{\zeta, \zeta'\} = \{1, j\}$ or $\{1, j^2\}$. In either case, $\rho(u)$ is a reflection. As G is generated by elements of 3, we obtain:

(11.2.8) $\rho(G) \simeq G$ *is a reflection group of rank* 2.

Because N' is the derived group of G, we have $\rho(N') \subseteq \mathsf{SL}_{\mathbb{C}}(V)$. As a consequence, $\mathsf{Ref}(\rho(G))$ consists of $\rho(u)$, where u belongs to the set of elements of order 3 in G. Hence

$$|\mathsf{Ref}(G)| = 8.$$

If we denote by (d_1, d_2) the sequence of degrees of $\rho(G)$ (see the Shephard-Todd-Chevalley theorem in Appendix C), then

$$d_1 + d_2 - 2 = |\mathsf{Ref}(\rho(G))| = 8 \quad \text{and} \quad d_1 d_2 = |G| = 24,$$

and so $d_1 = 4$ and $d_2 = 6$.

The group $\rho(G)$ is an exceptional complex reflection group denoted G_4 in the classification of Shephard and Todd [ShTo] (see also the recent book of Lehrer and Taylor [Leh]). The group G therefore admits a presentation $G =< a, b \mid a^3 = b^3 = 1, aba = bab >$, sending, for example, a to $\begin{pmatrix} 1 & 1 \\ 0 & 1 \end{pmatrix}$ and b to $\begin{pmatrix} 1 & 0 \\ -1 & 1 \end{pmatrix}$.

11.2.5. *The Group* $\mathsf{PSL}_2(\mathbb{F}_3)$ *and the Isometries of the Tetrahedron*

Denote by $\mathbb{Q}[\mathbf{P}^1(\mathbb{F}_3)]$ the representation over $\mathbb{Q}$ of G with character $\mathsf{Ind}_B^G 1_B$. Denote by $\mathscr{H}$ the hyperplane of $\mathbb{Q}[\mathbf{P}^1(\mathbb{F}_3)]$ equal to

$$\mathscr{H} = \{ \sum_{\delta \in \mathbf{P}^1(\mathbb{F}_3)} a_\delta \delta \mid \sum_{\delta \in \mathbf{P}^1(\mathbb{F}_3)} a_\delta = 0 \}.$$

Then $\mathscr{H}$ is G-stable and the corresponding representation $G \to \mathsf{GL}_{\mathbb{Q}}(\mathscr{H})$ admits St_G as character, and factorises to give an injection

$$\tau_3 : \mathsf{PSL}_2(\mathbb{F}_3) \hookrightarrow \mathsf{GL}_{\mathbb{Q}}(\mathscr{H}).$$

Let us equip $\mathbb{Q}[\mathbf{P}^1(\mathbb{F}_3)]$ with a scalar product such that the set $\mathbf{P}^1(\mathbb{F}_3)$ provides an orthonormal basis. Denote by $\pi : \mathbb{Q}[\mathbf{P}^1(\mathbb{F}_3)] \to \mathscr{H}$ the orthogonal projection and consider

$$\Delta = \{\pi(\delta) \mid \delta \in \mathbf{P}^1(\mathbb{F}_3)\}.$$

Then Δ is a regular tetrahedron in $\mathscr{H}$ upon which $\mathrm{PSL}_2(\mathbb{F}_3)$ acts (via τ_3) as isometries. One may easily verify that $\mathrm{PSL}_2(\mathbb{F}_3)$ is the group of oriented isometries of the tetrahedron Δ.

11.3. The Case when $q = 5$

Hypothesis. *In this and only this section we suppose that* $q = 5$.

11.3.1. Structure

We have

(11.3.1) $$|\mathrm{SL}_2(\mathbb{F}_5)| = 120, \quad |\mathrm{PGL}_2(\mathbb{F}_5)| = 120 \quad \text{and} \quad |\mathrm{PSL}_2(\mathbb{F}_5)| = 60.$$

By Theorem 1.2.4, $\mathrm{PSL}_2(\mathbb{F}_5)$ is simple (of order 60), therefore

(11.3.2) $$\mathrm{PSL}_2(\mathbb{F}_5) \simeq \mathfrak{A}_5.$$

This isomorphism may be seen as follows. Firstly, note that $|N| = 8$ and $|G| = 8 \times 15$, and therefore N is a Sylow 2-subgroup of G. As $q \equiv 5 \mod 8$, we have $|\mathrm{N}_G(N)/N| = 3$, which shows that $\mathrm{N}_G(N)$ is of index 5 in G. The action of G on $G/\mathrm{N}_G(N)$ (or, equivalently, the action of G on its five Sylow 2-subgroups) induces a morphism $G \to \mathfrak{S}_5$ having kernel Z and whose image, of order 60, is necessarily equal to the alternating group $\mathfrak{A}_5$. We can also see $\mathrm{N}_G(N)$ as being the image of $\mathrm{SL}_2(\mathbb{F}_3)$ under the morphism $\rho_{3\to 5}$ of Example 11.2.7.

It follows that $G = \mathrm{SL}_2(\mathbb{F}_5)$ is a non-trivial central extension of $\mathfrak{A}_5$ (which is unique up to isomorphism). Note that the isomorphism $\mathrm{PSL}_2(\mathbb{F}_5) \simeq \mathfrak{A}_5$ can be seen in other ways (see Exercise 11.1).

11.3.2. Character Table

The character table 5.4 specialises as follows. Let i denote an element of $\mathbb{F}_5^\times$ such that $i^2 = -1$, j an element of μ_6 such that $j^2 + j + 1 = 0$ (i.e. j is of order 3), $i^\wedge$ a linear character of $\mathbb{F}_5^\times$ of order 4, $j^\wedge$ a linear character of μ_6 of order 3 and denote by $-j^\wedge$ the linear character of order 6 of μ_6 equal to $\theta_0 j^\wedge$. Then, setting

$$\omega = \frac{1+\sqrt{5}}{2} \quad \text{and} \quad \omega^* = \frac{1-\sqrt{5}}{2},$$

the character table of $\mathrm{SL}_2(\mathbb{F}_5)$ is given in Table 11.2.

Table 11.2 Character table of $\mathrm{SL}_2(\mathbb{F}_5)$

g	I_2	$-I_2$	$d(i)$	$d'(j)$	$d'(-j)$	u_+	u_-	$-u_+$	$-u_-$
$\lvert \mathrm{Cl}_G(g)\rvert$	1	1	30	20	20	12	12	12	12
$o(g)$	1	2	4	3	6	5	5	10	10
$C_G(g)$	G	G	T	T'	T'	ZU	ZU	ZU	ZU
1_G	1	1	1	1	1	1	1	1	1
$R'_+(\theta_0)$	2	-2	0	-1	1	$-\omega^*$	$-\omega$	ω^*	ω
$R'_-(\theta_0)$	2	-2	0	-1	1	$-\omega$	$-\omega^*$	ω	ω^*
$R_+(\alpha_0)$	3	3	-1	0	0	ω	ω^*	ω	ω^*
$R_-(\alpha_0)$	3	3	-1	0	0	ω^*	ω	ω^*	ω
$R'(j^\wedge)$	4	4	0	1	1	-1	-1	-1	-1
$R'(-j^\wedge)$	4	-4	0	1	-1	1	1	-1	-1
St_G	5	5	1	-1	-1	0	0	0	0
$R(i^\wedge)$	6	-6	0	0	0	1	1	-1	-1

11.3.3. The Group $\mathrm{SL}_2(\mathbb{F}_5)$ *as a Subgroup of* $\mathrm{SL}_2(\mathbb{F}_{\ell^r})$

Let ζ_5 be the fifth root of unity in K equal to $\chi_+(1)$ (recall that the linear character $\chi_+ : \mathbb{F}_5^+ \to K^\times$ was fixed in Section 5.2). We then have, by definition of $\sqrt{5}$ (see §5.2.3),

$$\zeta_5 + \zeta_5^{-1} = \frac{-1+\sqrt{5}}{2}.$$

We remark that G admits two irreducible representations of degree 2 and in both representations the centre Z acts non-trivially. The following proposition shows that we can realise these representations over the field $\mathbb{Q}(\zeta_5)$.

Proposition 11.3.3. *There exists an (injective) homomorphism of groups* $\rho'_\pm$: $\mathrm{SL}_2(\mathbb{F}_5) \to \mathrm{GL}_2(\mathbb{Q}(\zeta_5))$ *with character* $R'_\pm(\theta_0)$.

Proof. Denote by m the Schur indicator of $R'_\pm(\theta_0)$ over $\mathbb{Q}(\zeta_5)$ (see [Isa, Definition 10.1]). The proposition amounts to showing that $m = 1$ (see [Isa, Corollary 10.2 (e)]). But $R'_\pm(\theta_0)$ appears with multiplicity 1 in $\mathrm{Ind}_U^G \chi_+$. As χ_+ has values in $\mathbb{Q}(\zeta_5)$ and has Schur index 1 over this field, it results from [Isa, Lemma 10.4] that $m = 1$. □

REMARK 11.3.4 – The field $\mathbb{Q}(\zeta_5)$ is the smallest field over which a representation of $\mathsf{SL}_2(\mathbb{F}_5)$ admits $\mathsf{R}'_{\pm}(\theta_0)$ as character. Indeed, it is impossible to realise such a representation over $\mathbb{Q}$, because the character has values in $\mathbb{Q}(\sqrt{5}) = \mathbb{Q}(\zeta_5 + \zeta_5^{-1})$. Nor is it possible to realise such a representation over $\mathbb{Q}(\sqrt{5})$ because this field is contained in $\mathbb{R}$ and the Frobenius-Schur indicator of $\mathsf{R}'_{\pm}(\theta_0)$ is -1 (as may be easily calculated). Now $\mathbb{Q}$ and $\mathbb{Q}(\sqrt{5})$ are the only proper subfields of $\mathbb{Q}(\zeta_5)$, and the minimality of $\mathbb{Q}(\zeta_5)$ follows. □

Because the ring $\mathbb{Z}[\zeta_5]$ is a principal ideal domain, it follows from Proposition 11.3.3 and B.3.1 that there exists a morphism $\rho'_{\pm} : \mathsf{SL}_2(\mathbb{F}_5) \to \mathsf{GL}_2(\mathbb{Z}[\zeta_5])$ such that $\mathsf{Tr}(\rho'_{\pm}(g)) = \mathsf{R}'_{\pm}(\theta_0)(g)$ for all $g \in \mathsf{SL}_2(\mathbb{F}_5)$. Because $\mathsf{SL}_2(\mathbb{F}_5)$ is perfect we have

$$\mathsf{Im}\,\rho'_{\pm} \subseteq \mathsf{SL}_2(\mathbb{Z}[\zeta_5]). \tag{11.3.5}$$

If ℓ is a prime number (possibly equal to $p = 5$) and if λ is a maximal ideal of $\mathbb{Z}[\zeta_5]$ containing ℓ, we denote by ρ'_{λ} the reduction modulo λ of ρ'_{+}:

$$\rho'_{\lambda} : \mathsf{SL}_2(\mathbb{F}_5) \longrightarrow \mathsf{SL}_2(\mathbb{Z}[\zeta_5]/\lambda).$$

Proposition 11.3.6. *The representation* $\rho'_{\lambda} : \mathsf{SL}_2(\mathbb{F}_5) \longrightarrow \mathsf{SL}_2(\mathbb{Z}[\zeta_5]/\lambda)$ *is (absolutely) irreducible. Moreover:*

(a) *If* $\ell = 2$, *then* $\mathsf{Ker}\,\rho'_{\lambda} = \{\pm \mathsf{I}_2\}$.
(b) *If* ℓ *is odd, then* ρ_{λ} *is injective.*

Proof. The only normal subgroups of $\mathsf{SL}_2(\mathbb{F}_5)$ are $\{\mathsf{I}_2\}$, $\{\pm\mathsf{I}_2\}$ and $\mathsf{SL}_2(\mathbb{F}_5)$. As $\mathsf{Tr}(\rho'_{\lambda}(\mathbf{d}'(j))) = 1 \neq 0$, we obtain that $\mathsf{Ker}\,\rho'_{\lambda} \neq \mathsf{SL}_2(\mathbb{F}_5)$.

On the other hand, $\rho'_{\lambda}(-\mathsf{I}_2) = \begin{pmatrix} -1 & 0 \\ 0 & -1 \end{pmatrix}$, which completes the proof of (a) and (b). Now, if the representation were not (absolutely) irreducible, the image of ρ'_{λ} would be contained, up to conjugation, in the subgroup of $\mathsf{SL}_2(\mathbb{F}_{\ell^r})$ formed by upper triangular matrices (for some $r \geqslant 1$). In particular, the image would be solvable, which is not the case. □

Denote by R_5 the ring of integers of $\mathbb{Q}(\sqrt{5})$ (recall that $R_5 = \mathbb{Z}[(1+\sqrt{5})/2] = \mathbb{Z}[\zeta_5 + \zeta_5^{-1}]$). As $\mathsf{R}'_{+}(\theta_0)$ takes values in R_5, we have $\mathsf{Tr}(\rho'_{\lambda}(g)) \in R_5/(\lambda \cap R_5)$ for all $g \in G$. Consequently, it results from [Isa, Theorem 9.14(b)] (with this result following from Wedderburn's theorem that all finite division rings are commutative) that, after conjugating by an element of $\mathsf{GL}_2(\mathbb{Z}[\zeta_5]/\lambda)$, we obtain an (absolutely) irreducible representation

$$\rho_{5\to\ell} : \mathsf{SL}_2(\mathbb{F}_5) \longrightarrow \mathsf{SL}_2(R_5/(\lambda \cap R_5)).$$

We can then deduce from Proposition 11.3.6 the following results.

Corollary 11.3.7. *With the above notation, we have:*

(a) *If $\ell = 2$, then $\lambda \cap R_5 = 2R_5$, $\operatorname{Ker}\rho_{5\to\ell} = \{\pm I_2\}$, $R_5/(\lambda \cap R_5) = \mathbb{F}_4$ and $\rho_{5\to\ell}$ induces an isomorphism $\mathrm{PSL}_2(\mathbb{F}_5) \simeq \mathrm{SL}_2(\mathbb{F}_4)$.*
(b) *If $\ell > 2$, then $\rho_{5\to\ell}$ is injective.*
(c) *If $\ell = 5$, then $\ell \cap R_5 = \sqrt{5}R_5$, $R_5/(\lambda \cap R_5) = \mathbb{F}_5$ and $\rho_{5\to\ell}$ is an isomorphism $\mathrm{SL}_2(\mathbb{F}_5) \simeq \mathrm{SL}_2(\mathbb{F}_5)$.*
(d) *If $\ell \equiv \pm 1 \mod 10$, then $R_5/(\lambda \cap R_5) = \mathbb{F}_\ell$ and therefore $\mathrm{SL}_2(\mathbb{F}_5)$ is isomorphic (via $\rho_{5\to\ell}$) to a subgroup of $\mathrm{SL}_2(\mathbb{F}_\ell)$.*
(e) *If $\ell \equiv \pm 3 \mod 10$, then $R_5/(\lambda \cap R_5) = \mathbb{F}_{\ell^2}$ and therefore $\mathrm{SL}_2(\mathbb{F}_5)$ is isomorphic (via $\rho_{5\to\ell}$) to a subgroup of $\mathrm{SL}_2(\mathbb{F}_{\ell^2})$.*

Proof. It is enough to determine the structure of the field $R_5/(\lambda \cap R_5)$, that is to say, to decide whether $X^2 - X - 1$ (the minimal polynomial of $(1+\sqrt{5})/2$) admits a root in $\mathbb{F}_\ell$ or not. If $\ell \in \{2,5\}$ this is straightforward, and if $\ell \notin \{2,5\}$ we must determine whether 5 is a square modulo ℓ. By the law of quadratic reciprocity, this is equivalent to determining if ℓ is a square modulo 5. The statements of the corollary then follow. □

EXAMPLE 11.3.8 – Proposition 11.3.7(e) shows that $\mathrm{SL}_2(\mathbb{F}_5)$ is isomorphic to a subgroup H of $\mathrm{SL}_2(\mathbb{F}_9)$. As H is of index 6, the left action of $\mathrm{SL}_2(\mathbb{F}_9)$ on $\mathrm{SL}_2(\mathbb{F}_9)/H$ induces a homomorphism $\mathrm{SL}_2(\mathbb{F}_9) \to \mathfrak{S}_6$. As $\mathrm{SL}_2(\mathbb{F}_9)$ is equal to its derived group and $\mathrm{PSL}_2(\mathbb{F}_9)$ is simple (see Theorem 1.2.4), one may compare cardinalities to conclude that the above morphism induces an isomorphism $\mathrm{PSL}_2(\mathbb{F}_9) \xrightarrow{\sim} \mathfrak{A}_6$. □

EXAMPLE 11.3.9 – Proposition 11.3.7(d) shows that $\mathrm{SL}_2(\mathbb{F}_5)$ is isomorphic to a subgroup H of $\mathrm{SL}_2(\mathbb{F}_{11})$. As H is of index 11, the left action of $\mathrm{SL}_2(\mathbb{F}_{11})$ on $\mathrm{SL}_2(\mathbb{F}_{11})/H$ yields a homomorphism $\mathrm{SL}_2(\mathbb{F}_{11}) \to \mathfrak{S}_{11}$. As $\mathrm{SL}_2(\mathbb{F}_{11})$ is equal to its derived group and $\mathrm{PSL}_2(\mathbb{F}_{11})$ is simple (see Theorem 1.2.4), this morphism induces an injection $\mathrm{PSL}_2(\mathbb{F}_{11}) \hookrightarrow \mathfrak{A}_{11}$ with image a transitive subgroup. This morphism is "exceptional" by comparison with the natural action of $\mathrm{PSL}_2(\mathbb{F}_{11})$ on the set of 12 elements of $\mathbf{P}^1(\mathbb{F}_{11})$. One may show that the image of $\mathrm{PSL}_2(\mathbb{F}_{11}) \hookrightarrow \mathfrak{A}_{11}$ is contained in the Mathieu group M_{11} (a simple subgroup of $\mathfrak{A}_{11}$ of order 7920). This image, of index 12, then allows one to construct a transitive action of the Mathieu group M_{11} on a set of order 12. □

11.3.4. *The Group* $\mathrm{SL}_2(\mathbb{F}_5) \times \mathbb{Z}/5\mathbb{Z}$ *as a Reflection Group of Rank* 2

Let us reconsider the irreducible representation $\rho'_+ : \mathrm{SL}_2(\mathbb{F}_5) \to \mathrm{SL}_2(\mathbb{Q}(\zeta_5))$ and denote by $\mu_5^{\mathbb{C}}$ the subgroup of $\mathrm{GL}_2(\mathbb{Q}(\zeta_5))$ given by scalar transformations of order 1 or 5. Set $W^+ = \operatorname{Im}\rho'_+$. We have $\mu_5^{\mathbb{C}} \cap W^+ = \{1\}$. We may therefore consider

$$W = \mu_5^{\mathbb{C}} \times W^+ \subseteq \mathrm{GL}_2(\mathbb{Q}(\zeta_5)).$$

We have

(11.3.10) $$\mathrm{D}(W) = W^+.$$

Proposition 11.3.11. *W is a subgroup of $\mathrm{GL}_2(\mathbb{Q}(\zeta_5))$ generated by reflections. All its reflections are of order 5 and $|\mathrm{Ref}(W)| = 48$. Furthermore, $\mathrm{D}(W) = W^+$ and $\mathrm{Z}(W) = \mu_{10}$. The sequence of degrees of W is $(20, 30)$ and W is the group denoted G_{16} in the Shephard-Todd classification.*

Proof. Amongst the elements of order 5 of W^+, denote by $\mathscr{E}_5^+$ the set of elements whose eigenvalues are ζ_5^2 and ζ_5^{-2} (the others have eigenvalues ζ_5 and ζ_5^{-1}). Set

$$\mathscr{R}_5 = \{\zeta_5^{-2} g \mid g \in \mathscr{E}_5^+\}.$$

Then the elements of $\mathscr{R}_5$ are reflections of order 5 (with eigenvalues 1 and ζ_5, and determinant ζ_5). We will show that

$(*)$ $$W \text{ is generated by } \mathscr{R}_5.$$

Indeed, if W' denotes the subgroup of W generated by $\mathscr{R}_5$ and if $\pi_+ \colon W \to W^+$ denotes the canonical projection, then $\pi_+(W')$ is the subgroup of W^+ generated by $\mathscr{E}_5^+$. As $\mathscr{E}_5^+$ is stable under conjugation, $\pi_+(W')$ is a normal subgroup of W^+ generated by elements of order 5, and hence is equal to $W^+ \simeq \mathrm{SL}_2(\mathbb{F}_5)$ (see Theorem 1.2.4).

In particular, 120 divides $|W'|$, and W' is of index dividing 5 in W. On the other hand the homomorphism $\det \colon W' \to \mu_5$ is surjective, with kernel $W^+ \cap W'$. Therefore $W^+ \cap W'$ is a normal subgroup of W^+ of index dividing 5. Hence $W^+ \cap W' = W^+$ by Theorem 1.2.4, that is to say $W^+ \subseteq W'$ and $|W'/W^+| = 5$. Hence $W = W'$, which shows $(*)$. The equalities $\mathrm{D}(W) = W^+$ and $\mathrm{Z}(W) = \mu_{10}$ now follow easily. Note that the order of a reflection is equal to the order of its determinant, which shows that all reflections are of order 5.

Now consider $f \colon \mathrm{Ref}(W) \to W^+$, $g \mapsto \det(g)^2 g$. The image f is indeed contained in W^+ as $\det(g)^5 = 1$. On the other hand, if $f(g) = f(g')$, then $g = \lambda g'$ with $\lambda \in \mu_5$, which is only possible if $\lambda = 1$ or $\lambda = \det(g)^{-1}$ (because g and g' are reflections of order 5). Hence the fibres of f have cardinality at most 2. Moreover, the image of f is contained in the set $\mathscr{E}_5$ of elements of order 5 of W^+. We conclude that, $|\mathrm{Ref}(W)| \leqslant 2 \cdot |\mathscr{E}_5| = 48$.

If we denote by (d_1, d_2) the sequence of degrees of W, then, by the Shephard-Todd-Chevalley theorem (see Appendix C), we have

$$d_1 d_2 = |W| = 600, \quad d_1 + d_2 - 2 \leqslant 48 \quad \text{and} \quad 10 \text{ divides } d_1 \text{ and } d_2.$$

This forces $(d_1, d_2) = (20, 30)$ and $|\mathrm{Ref}(W)| = 48$, as claimed. □

EXAMPLE 11.3.12 – In an analogous manner we can construct other reflection groups having W^+ as derived group. Let us view W^+ as a subgroup of $SL_2(\mathbb{Q}(\mu_{60}))$. Given an *even* divisor n of 60 let

$$W_n = W^+ \cdot \mu_n.$$

Then one can show, by similar arguments to those employed in the proof of Proposition 11.3.11 (with some modifications ...), that

W_n is generated by reflections if and only if $n \in \{4, 6, 10, 12, 20, 30, 60\}$.

If this is the case, denote by $(d_1^{(n)}, d_2^{(n)})$ the sequence of degrees. These degrees together with the name of W_n in the Shephard-Todd classification are given in Table 11.3. □

Table 11.3 Reflection groups having $SL_2(\mathbb{F}_5)$ as derived group

n	$\|W_n\|$	$\|\mathrm{Ref}(W_n)\|$	$(d_1^{(n)}, d_2^{(n)})$	$Z(W_n)$	Standard name
4	240	30	(12, 20)	μ_4	G_{22}
6	360	40	(12, 30)	μ_6	G_{20}
10	600	48	(20, 30)	μ_{10}	G_{16}
12	720	70	(12, 60)	μ_{12}	G_{21}
20	1200	78	(20, 60)	μ_{20}	G_{17}
30	1800	88	(30, 60)	μ_{30}	G_{18}
60	3600	118	(60, 60)	μ_{60}	G_{19}

11.3.5. *The Group* $PSL_2(\mathbb{F}_5)$, *the Dodecahedron and the Icosahedron*

The two irreducible representations of $SL_2(\mathbb{F}_5)$ of degree 3 can be realised over the real numbers, as may be easily shown by calculating the Frobenius-Schur indicator (which is 1 in both cases). Both representations contain $-I_2$ in their kernel, and therefore factorise to yield a representation of $PSL_2(\mathbb{F}_5)$ with image contained in the special linear group (as $SL_2(\mathbb{F}_5)$ is equal to its derived group). Let us choose a homomorphism

$$\rho \colon \mathsf{PSL}_2(\mathbb{F}_5) \longrightarrow \mathsf{SL}_{\mathbb{R}}(V),$$

where V is an $\mathbb{R}$-vector space of dimension 3. We may fix an invariant scalar product $\langle , \rangle$ on V, so that ρ gives a homomorphism to the group of oriented isometries

$$\rho \colon \mathsf{PSL}_2(\mathbb{F}_5) \longrightarrow \mathsf{SO}_{\mathbb{R}}(V, \langle , \rangle).$$

Denote by $\mathscr{E}_3$ (respectively $\mathscr{E}_3$) the set of elements of order 3 (respectively 5) of $\mathsf{PSL}_2(\mathbb{F}_5)$. We have

$$|\mathscr{E}_3| = 20 \quad \text{and} \quad |\mathscr{E}_5| = 24.$$

Every non-trivial element of $\mathsf{SO}_{\mathbb{R}}(V, \langle , \rangle)$ is the rotation around a well-defined axis, and therefore the intersection of this axis with the unit sphere $\mathscr{S}$ consists of two elements. Let

$$\Delta_3 = \{v \in \mathscr{S} \mid \exists\, g \in \mathscr{E}_3,\ g \cdot v = v\}$$

and
$$\Delta_5 = \{v \in \mathscr{S} \mid \exists\, g \in \mathscr{E}_5,\ g \cdot v = v\}.$$

If $v \in \Delta_3$ (respectively $v \in \Delta_5$), there exist two (respectively four) elements of $\mathscr{E}_3$ (respectively $\mathscr{E}_5$) which fix it. In particular,

$$|\Delta_3| = \frac{2 \cdot 20}{2} = 20 \quad \text{and} \quad |\Delta_5| = \frac{2 \cdot 24}{4} = 12.$$

Proposition 11.3.13. *The group* $\mathsf{PSL}_2(\mathbb{F}_5)$ *acts transitively on* Δ_3 *and* Δ_5.

Proof. Let $v \in \Delta_3$. Denote by H the stabiliser, in $\mathsf{PSL}_2(\mathbb{F}_5)$, of v and let $n = |H|$. Then H is isomorphic to a finite subgroup of the group of oriented isometries of the orthogonal to v, which is of dimension 2. Hence H is cyclic. Moreover, by construction, 3 divides n. The possible orders of the elements of $\mathsf{PSL}_2(\mathbb{F}_5)$ are 1, 2, 3, 4 or 5 and hence $n = 3$. Consequently, the orbit of v under $\mathsf{PSL}_2(\mathbb{F}_5)$ has cardinality $20 = |\Delta_3|$. The result follows.

The transitivity of the action on Δ_5 is shown in the same way. □

From Proposition 11.3.13 we may easily conclude the following corollary.

Corollary 11.3.14. Δ_3 *(respectively* Δ_5*) is a regular dodecahedron (respectively an icosahedron) and* $\mathsf{PSL}_2(\mathbb{F}_5)$ *is its group of oriented isometries.*

11.4. The Case when $q = 7$

Hypothesis. *In this and only this section we suppose that* $q = 7$.

11.4.1. *Structure*

We have

(11.4.1) $$|\mathrm{SL}_2(\mathbb{F}_7)| = 336, \quad |\mathrm{PGL}_2(\mathbb{F}_7)| = 336 \quad \text{and} \quad |\mathrm{PSL}_2(\mathbb{F}_7)| = 168.$$

Recall that $\mathrm{PSL}_2(\mathbb{F}_7)$ is simple. However, contrary to the case of the groups $\mathrm{PSL}_2(\mathbb{F}_3)$ and $\mathrm{PSL}_2(\mathbb{F}_5)$, $\mathrm{PSL}_2(\mathbb{F}_7)$ is not isomorphic to an alternating group. We will show below (see Proposition 11.4.4) that

(11.4.2) $$\mathrm{PSL}_2(\mathbb{F}_7) \simeq \mathrm{GL}_3(\mathbb{F}_2).$$

11.4.2. *Character Table*

The character table 5.4 specialises as follows. Let j denote an element of μ_6 of order 3, $j^\wedge$ a linear character of order 3 of μ_6, i a square root of -1 in μ_8, $i^\wedge$ a linear character of order 4 of μ_8, ζ_8 an element of μ_8 of order 8, $\zeta_8^\wedge$ a linear character of μ_8 of order 8 and $\sqrt{2} = \zeta_8^\wedge(\zeta_8) + \zeta_8^\wedge(\zeta_8)^{-1}$. We let

$$\varpi = \frac{1+\sqrt{-7}}{2} \quad \text{and} \quad \varpi^* = \frac{1-\sqrt{-7}}{2}.$$

The character table of $\mathrm{SL}_2(\mathbb{F}_7)$ is given in Table 11.4.

11.4.3. *The Isomorphism Between the Groups* $\mathrm{PSL}_2(\mathbb{F}_7)$ *and* $\mathrm{GL}_3(\mathbb{F}_2)$.

By Table 11.4, the group $\mathrm{SL}_2(\mathbb{F}_7)$ admits two irreducible representations of dimension 3 over K on which $-\mathrm{I}_2$ acts trivially. These representations may be realised in the ℓ-adic cohomology of the variety $\mathbf{Y}$ (with coefficients in a sufficiently large extension of $\mathbb{Q}_\ell$) for all $\ell \neq 7$. In particular, we can take $\ell = 2$, so that K is a finite extension of $\mathbb{Q}_2$. Denote by $\mathcal{O}$ the ring of integers of K. Proposition B.3.1, together with the fact that $\mathcal{O}$ is a principal ideal domain, shows that there exists an injective homomorphism

$$\rho \colon \mathrm{PSL}_2(\mathbb{F}_7) \hookrightarrow \mathrm{GL}_3(\mathcal{O})$$

such that

(11.4.3) $$\mathrm{Tr}(\rho(\bar{g})) = \mathrm{R}'_+(\theta_0)(g)$$

Table 11.4 Character table of $\mathrm{SL}_2(\mathbb{F}_7)$

g	I_2	$-\mathrm{I}_2$	$\mathbf{d}(j)$	$\mathbf{d}(-j)$	$\mathbf{d}'(i)$	$\mathbf{d}'(\zeta_8)$	$\mathbf{d}'(\zeta_8^3)$	u_+	u_-	$-u_+$	$-u_-$
$\lvert \mathrm{Cl}_G(g)\rvert$	1	1	56	56	42	42	42	24	24	24	24
$o(g)$	1	2	3	6	4	8	8	7	7	14	14
$\mathrm{C}_G(g)$	G	G	T	T	T'	T'	T'	ZU	ZU	ZU	ZU
1_G	1	1	1	1	1	1	1	1	1	1	1
$\mathrm{R}'_+(\theta_0)$	3	3	0	0	-1	1	1	$-\varpi^*$	$-\varpi$	$-\varpi^*$	$-\varpi$
$\mathrm{R}'_-(\theta_0)$	3	3	0	0	-1	1	1	$-\varpi$	$-\varpi^*$	$-\varpi$	$-\varpi^*$
$\mathrm{R}_+(\alpha_0)$	4	-4	1	-1	0	0	0	ϖ	ϖ^*	$-\varpi$	$-\varpi^*$
$\mathrm{R}_-(\alpha_0)$	4	-4	1	-1	0	0	0	ϖ^*	ϖ	$-\varpi^*$	$-\varpi$
$\mathrm{R}'(i^\wedge)$	6	6	0	0	2	0	0	-1	-1	-1	-1
$\mathrm{R}'(\zeta_8^\wedge)$	6	-6	0	0	0	$\sqrt{2}$	$-\sqrt{2}$	-1	-1	1	1
$\mathrm{R}'(\zeta_8^{\wedge 3})$	6	-6	0	0	0	$-\sqrt{2}$	$\sqrt{2}$	-1	-1	1	1
St_G	7	7	1	1	-1	-1	-1	0	0	0	0
$\mathrm{R}(j^\wedge)$	8	8	-1	-1	0	0	0	1	1	1	1
$\mathrm{R}(j^\wedge)$	8	-8	-1	1	0	0	0	1	1	-1	-1

for all $g \in \mathrm{SL}_2(\mathbb{F}_7)$ (we denote by $\bar{g}$ the image of g in $\mathrm{PSL}_2(\mathbb{F}_7)$). Recall that $\mathfrak{l}$ denotes the maximal ideal of $\mathscr{O}$. We denote by

$$\rho_{\mathfrak{l}} : \mathrm{PSL}_2(\mathbb{F}_7) \to \mathrm{GL}_3(\mathscr{O}/\mathfrak{l})$$

the reduction modulo $\mathfrak{l}$ of ρ.

Now $(1+\sqrt{-7})/2$ is a root of the polynomial $X^2 - X + 2$ which splits modulo 2 and therefore $(1+\sqrt{-7})/2 \in \mathbb{Z}_2$. We conclude from 11.4.3 and the character table 11.4 that $\mathrm{Tr}(\rho_{\mathfrak{l}}(g)) \in \mathbb{F}_2$ for all $g \in \mathrm{PSL}_2(\mathbb{F}_7)$. Furthermore, by [Isa, Theorem 9.14(b)], after conjugating $\rho_{\mathfrak{l}}$ by a suitable element of $\mathrm{GL}_3(\mathscr{O}/\lambda)$, we obtain a homomorphism

$$\rho_{7\to 2} : \mathrm{PSL}_2(\mathbb{F}_7) \longrightarrow \mathrm{GL}_3(\mathbb{F}_2).$$

Proposition 11.4.4. *The homomorphism $\rho_{7\to 2}\colon \mathsf{PSL}_2(\mathbb{F}_7) \longrightarrow \mathsf{GL}_3(\mathbb{F}_2)$ is an isomorphism of groups.*

Proof. As $|\mathsf{PSL}_2(\mathbb{F}_7)| = |\mathsf{GL}_3(\mathbb{F}_2)| = 168$, it is enough to show that $\rho_{7\to 2}$ is injective. Because $\mathsf{PSL}_2(\mathbb{F}_7)$ is simple (see Theorem 1.2.4), it is enough to show that $\mathsf{Ker}\,\rho_{7\to 2} \neq \mathsf{PSL}_2(\mathbb{F}_7)$. But, by 11.4.3 and the character table 11.4, there exists an element $g \in \mathsf{PSL}_2(\mathbb{F}_7)$ such that $\mathsf{Tr}(\rho_{7\to 2}(g)) = 0$. Therefore $\rho_{7\to 2}(g)$ is not the identify matrix, which completes the proof of the proposition. □

REMARK 11.4.5 – The group $\mathsf{GL}_3(\mathbb{F}_2)$ has a "natural" action on the set of 7 elements given by the projective plane $\mathbf{P}^2(\mathbb{F}_2)$. On the other hand, from the point of view of $\mathsf{PSL}_2(\mathbb{F}_7)$, the most "natural" action is on the set of eight elements given by the projective line $\mathbf{P}^1(\mathbb{F}_7)$. □

11.4.4. *The Group* $\mathsf{PSL}_2(\mathbb{F}_7) \times \mathbb{Z}/2\mathbb{Z}$ *as a Reflection Group of Rank* 3

Let us fix an irreducible representation $\rho\colon \mathsf{PSL}_2(\mathbb{F}_7) \to \mathsf{GL}_3(\mathbb{C})$ with character $\mathsf{R}'_+(\theta_0)$. Denote by W^+ the image of ρ. As ρ is injective and $\mathsf{PSL}_2(\mathbb{F}_7)$ is equal to its derived group, we have $W^+ \subseteq \mathsf{SL}_3(\mathbb{C})$. Denote by $\mu_2^{\mathbb{C}}$ the subgroup of $\mathsf{GL}_3(\mathbb{C})$ given by scalar transformations of order 1 or 2 and set

$$W = W^+ \times \mu_2^{\mathbb{C}} \simeq \mathsf{PSL}_2(\mathbb{F}_7) \times \mathbb{Z}/2\mathbb{Z}.$$

Then

$$(11.4.6) \qquad |W| = 336.$$

Let $\mathscr{I}$ denote the set of elements of W^+ of order 2. We have

$$(11.4.7) \qquad |\mathscr{I}| = 21.$$

Set $\mathscr{R} = \{-s \mid s \in \mathscr{I}\}$. Then

$$(11.4.8) \qquad |\mathscr{R}| = 21.$$

Proposition 11.4.9. *The elements of $\mathscr{R}$ are reflections, the group W is generated by $\mathscr{R}$ and $\mathscr{R} = \mathsf{Ref}(W)$. The sequence of degrees of W is $(4, 6, 14)$. The group W is isomorphic to the complex reflection group G_{24} in the Shephard-Todd classification.*

Proof. An element of order 2 in $\mathsf{SL}_3(\mathbb{C})$ necessarily has characteristic polynomial $(X+1)^2(X-1)$. Hence, if we multiply such an element by -1 we obtain a reflection (of order 2). Hence the elements of $\mathscr{R}$ are reflections.

Let $W' = \langle \mathscr{R} \rangle$. Denote by $\pi\colon W \to W^+$ the canonical projection. Then $\pi(W')$ is a subgroup of W^+ containing $\mathscr{I}$ and is therefore normal and non-trivial. The simplicity of $\mathrm{PSL}_2(\mathbb{F}_7)$ implies that $\pi(W') = W^+$. Hence $|W'| \in \{168, 336\}$. Now, if $|W'| = 168$, then $W' \simeq \mathrm{PSL}_2(\mathbb{F}_7)$ and therefore the morphism det is non-trivial on $\mathrm{PSL}_2(\mathbb{F}_7)$, which is impossible. Hence $|W'| = 336$, and $\mathscr{R}$ generates W.

The group W is therefore generated by its reflections and $\mathscr{R} \subseteq \mathrm{Ref}(W)$. On the other hand, if $s \in \mathrm{Ref}(W)$, then $\det(s) = -1$, hence $-s \in W^+$ and $-s$ is an involution. We have therefore shown that $\mathrm{Ref}(W) = \mathscr{R}$.

Now, denote by (d_1, d_2, d_3) the sequence of degrees of W. Then, by the Shephard-Todd-Chevalley theorem, d_1, d_2 and d_3 are even (as $|\mathrm{Z}(W)| = 2$), $d_1 + d_2 + d_3 - 3 = |\mathrm{Ref}(W)| = 21$ and $d_1 d_2 d_3 = |W| = 336$. It is straightforward to see that this forces $(d_1, d_2, d_3) = (4, 6, 14)$.

The last assertion is easily verified. □

Let us identify the algebra of polynomial functions on $\mathbb{C}^3$ with $\mathbb{C}[X, Y, Z]$. Proposition 11.4.9 shows that there exists a homogeneous polynomial $P \in \mathbb{C}[X, Y, Z]^W$ of degree 4. This homogeneous polynomial defines a plane projective curve

$$\mathbf{C} = \{[x; y; z] \in \mathbf{P}^2(\mathbb{C}) \mid P(x, y, z) = 0\}.$$

One may verify that this plane projective curve is smooth. As the curve is of degree $d = 4$, it has genus $g = (d-1)(d-2)/2 = 3$ [Har, Exercise I.7.2(b) and Proposition IV.1.1]. Now, by a theorem of Hurwitz [Har, Exercise IV.2.5], we have

$$|\mathrm{Aut}\,\mathbf{C}| \leqslant 84 \cdot (g-1) = 168.$$

As the group $\mathrm{PSL}_2(\mathbb{F}_7)$ acts faithfully on on this curve, we deduce that

$$\mathrm{Aut}\,\mathbf{C} \simeq \mathrm{PSL}_2(\mathbb{F}_7) \simeq \mathrm{GL}_3(\mathbb{F}_2). \tag{11.4.10}$$

Hence this curve provides an example in which the bound of Hurwitz is attained. One may verify that in fact $\mathbf{C}$ is the *Klein quartic* (defined, for example, in [Har, Exercise IV.5.7]).

Exercises

11.1. Suppose that $q = 5$. After numbering from 1 to 6 the points of $\mathbf{P}^1(\mathbb{F}_5)$ the action of $\widetilde{G}$ on $\mathbf{P}^1(\mathbb{F}_5)$ induces a homomorphism $\widetilde{G} \to \mathfrak{S}_6$, which we denote by ϕ. Set $\Gamma = \mathrm{Im}\,\phi \simeq \mathrm{PGL}_2(\mathbb{F}_5)$.

(a) Let n be a non-zero natural number and let H be a subgroup of $\mathfrak{S}_n$ of index n. Denote by $\phi_0 : \mathfrak{S}_n \to \mathfrak{S}_n$ the morphism induced by the action of $\mathfrak{S}_n$ on $\mathfrak{S}_n/H$ (after enumerating the elements of $\mathfrak{S}_n/H$ from 1 up to n).

(a1) Show that ϕ_0 is an automorphism. (***Hint***: If $n \neq 4$, use the simplicity of $\mathfrak{A}_n$ and for $n = 4$, note that $\mathfrak{A}_4$ does not have a subgroup of order 6.)

(a2) Show that, if n is the number corresponding to H in $\mathfrak{S}_n/H$, then $\phi_0(H)$ is the stabiliser of n in $\mathfrak{S}_n$. Deduce that $H \simeq \mathfrak{S}_{n-1}$.

(a3) Show that if H is a transitive subgroup of $\mathfrak{S}_n$, then ϕ_0 is not an inner automorphism.

(b) Deduce from (b) that $\Gamma \simeq \mathfrak{S}_5$ and that $\mathfrak{S}_6$ has a non-inner automorphism.

(c) Deduce a new proof that $\mathrm{PSL}_2(\mathbb{F}_5) \simeq \mathfrak{A}_5$ (see 11.3.2).

11.2* Let Γ be a finite group. Show that:

(a) If Γ is simple of order 60, then $\Gamma \simeq \mathrm{PSL}_2(\mathbb{F}_5) \simeq \mathfrak{A}_5$.

(b*) If Γ is simple of order 168, then $\Gamma \simeq \mathrm{PSL}_2(\mathbb{F}_7) \simeq \mathrm{GL}_3(\mathbb{F}_2)$.

(c*) If Γ is simple of order 360, then $\Gamma \simeq \mathrm{PSL}_2(\mathbb{F}_9) \simeq \mathfrak{A}_6$.

(d**) If Γ is simple of order 504, then $\Gamma \simeq \mathrm{PSL}_2(\mathbb{F}_8) = \mathrm{SL}_2(\mathbb{F}_8)$.

(e**) If Γ is simple of order 660, then $\Gamma \simeq \mathrm{PSL}_2(\mathbb{F}_{11})$.

(f*) If Γ is simple of order $\leqslant 1000$, then Γ is isomorphic to $\mathrm{PSL}_2(\mathbb{F}_q)$, for some $q \in \{5, 7, 8, 9, 11\}$.

11.3. Consider

$$\begin{aligned} \tau\colon \mathrm{SL}_2(\mathbb{F}_3) &\longrightarrow \mathbb{F}_3^+ \\ \begin{pmatrix} a & b \\ c & d \end{pmatrix} &\longmapsto (a^2+c^2)(ab+cd). \end{aligned}$$

Show that τ is a homomorphism of groups and that, if $x \in \mathbb{F}_3$, then $\tau(\mathbf{u}(x)) = x$. Show also that $N' = \operatorname{Ker}\tau$.

11.4. Let Q_8 be the quaternionic group of order 8 and let $\mathbb{K}$ be a field of characteristic zero. Suppose that there exists an irreducible representation of Q_8 over $\mathbb{K}$ of dimension 2. Show that -1 is a sum of two squares in $\mathbb{K}$. (***Hint:*** Set $Q_8 = \{\pm 1, \pm I, \pm J, \pm K\}$ (using standard notation) and fix an irreducible representation $\rho\colon Q_8 \to \mathrm{GL}_2(\mathbb{K})$. Show that we may always suppose that $\rho(I) = \begin{pmatrix} 0 & -1 \\ 1 & 0 \end{pmatrix}$ and that, if we set $\rho(J) = \begin{pmatrix} a & b \\ c & d \end{pmatrix}$, then the relations $\rho(J)^2 = \rho(I)^2$ and $(\rho(I)\rho(J))^2 = \rho(I)^2$ force $d = -a$, $b = c$ and $a^2 + b^2 = -1$.)

REMARK – Denote by $\mathbb{H}$ the "standard" algebra of quaternions over a field $\mathbb{K}$ of characteristic different from 2. That is $\mathbb{H} = \mathbb{K} \oplus \mathbb{K}I \oplus \mathbb{K}J \oplus \mathbb{K}K$, where $I^2 = J^2 = K^2 = -1$, $IJ = -JI = K$, $JK = -KJ = I$ and $KI = -IK = J$. Then $\mathbb{H}$ is *split* over $\mathbb{K}$ (that is, isomorphic to $\mathrm{Mat}_2(\mathbb{K})$) if and only if -1 is a sum of two squares in $\mathbb{K}$. This result appears similar to that shown in Exercise 11.4 above. In fact, the two results are equivalent if we view the algebra $\mathbb{H}$ as a quotient of the group algebra $\mathbb{K}Q_8$ by the ideal generated by $z+1$, where z is the unique non-trivial central element of Q_8.

11.5* Let $\mathbb{K}$ be a field of characteristic different from 2. Show that $\mathrm{SL}_2(\mathbb{F}_5)$ admits an irreducible faithful representation of dimension 2 over $\mathbb{K}$ if and only if 5 is a square in $\mathbb{K}$ and -1 is a sum of two squares in $\mathbb{K}$.

11.6 (McKay correspondence). Let Γ be a finite subgroup of $\mathsf{SL}_2(\mathbb{C})$ and χ_{nat} the character of the natural representation $\Gamma \hookrightarrow \mathsf{SL}_2(\mathbb{G})$. Define a graph as follows: the set of vertices of $\tilde{\mathscr{G}}_\Gamma$ is $\mathsf{Irr}\,\Gamma$ and two characters χ and χ' are connected by an edge if χ is an irreducible factor of $\chi'\chi_{\mathsf{nat}}$. We denote by $\mathscr{G}_\Gamma$ the full subgraph of $\tilde{\mathscr{G}}_\Gamma$ obtained by removing the vertex 1_Γ.

(a) Show that χ_{nat} is real valued (use Exercise 1.13 to show that $\chi_{\mathsf{nat}} = \chi_{\mathsf{nat}}^*$).
(b) Deduce that, if $\chi, \chi' \in \mathsf{Irr}\,\Gamma$, then χ is an irreducible factor of $\chi'\chi_{\mathsf{nat}}$ if and only if χ' is an irreducible factor of $\chi\chi_{\mathsf{nat}}$.

Denote by $\widetilde{\mathfrak{A}}_4$ (respectively $\widetilde{\mathfrak{A}}_5$) the subgroup of $\mathsf{SL}_2(\mathbb{C})$ given by the image of $\mathsf{SL}_2(\mathbb{F}_3)$ (respectively $\mathsf{SL}_2(\mathbb{F}_5)$) under the irreducible representation of dimension 2 with character $\mathsf{R}'(i^\wedge)$ (respectively $\mathsf{R}'_+(\theta_0)$). Denote by $\widetilde{\mathfrak{S}}_4$ the normaliser of $\widetilde{\mathfrak{A}}_4$ in $\mathsf{SL}_2(\mathbb{C})$. For background on Dynkin diagrams and root systems, see [Bou, Chapter VI].

(c) Calculate the character table of $\widetilde{\mathfrak{S}}_4$.
(d) Show that $\tilde{\mathscr{G}}_{\widetilde{\mathfrak{A}}_4}$ (respectively $\tilde{\mathscr{G}}_{\widetilde{\mathfrak{S}}_4}$, respectively $\tilde{\mathscr{G}}_{\widetilde{\mathfrak{A}}_5}$) is a graph of type $\tilde{E}_6$ (respectively $\tilde{E}_7$, respectively $\tilde{E}_8$). Here, $\tilde{E}_?$ denotes the affine Dynkin diagram associated to the Dynkin diagram of type $E_?$.
(e) Show that $\mathscr{G}_{\widetilde{\mathfrak{A}}_4}$ (respectively $\mathscr{G}_{\widetilde{\mathfrak{S}}_4}$, respectively $\mathscr{G}_{\widetilde{\mathfrak{A}}_5}$) is a Dynkin diagram of type E_6 (respectively E_7, respectively E_8).
(f) Generalise the above to all finite subgroups of $\mathsf{SL}_2(\mathbb{C})$.

11.7. Show that the group of outer automorphisms of $\mathfrak{A}_6$ is isomorphic to $\mathbb{Z}/2\mathbb{Z} \times \mathbb{Z}/2\mathbb{Z}$. (***Hint:*** Use Example 11.3.8 and Exercise 1.14.)

Chapter 12
Deligne-Lusztig Theory: an Overview*

This chapter gives a very succinct overview of Deligne-Lusztig theory. We will recall some of the principal results of the theory (including the parametrisation of characters and partition into blocks) with the goal of connecting this general theory with what we have seen for $\mathrm{SL}_2(\mathbb{F}_q)$. This chapter requires some knowledge of algebraic groups, for which we refer the reader to [Bor] or [DiMi].

For more details on the subjects covered in this chapter the reader is referred to the books [Lu1], [Carter], [DiMi] or [CaEn]. We also recommend the magnificent and foundational article on the subject, written by Deligne and Lusztig in 1976 [DeLu].

Hypotheses and notation. *In this and only this chapter we fix a reductive group* $\mathbf{G}$ *defined over a field* $\mathbb{F}$ *as well as a Frobenius endomorphism* $F \colon \mathbf{G} \to \mathbf{G}$ *which equips* $\mathbf{G}$ *with a rational structure over the finite field* $\mathbb{F}_q$. *Denote by* $\mathbf{B}$ *an* F*-stable Borel subgroup of* $\mathbf{G}$, $\mathbf{T}$ *an* F*-stable maximal torus of* $\mathbf{B}$ *and* $\mathbf{U}$ *the unipotent radical of* $\mathbf{B}$. *The Weyl group of* $\mathbf{G}$ *relative to* $\mathbf{T}$ *will be denoted* W; *by definition,* $W = \mathrm{N}_{\mathbf{G}}(\mathbf{T})$.

Furthermore, we return to the assumption that ℓ *is a prime number different from* p.

The *finite reductive group* is the group $\mathbf{G}^F$ of fixed points of F over $\mathbf{G}$:

$$\mathbf{G}^F = \{g \in \mathbf{G} \mid F(g) = g\}.$$

The object of Deligne-Lusztig theory is the study of ordinary and modular representations (in unequal characteristic) of the finite group $\mathbf{G}^F$. Recall one of the fundamental theorems concerning algebraic groups defined over a finite field.

C. Bonnafé, *Representations of* $\mathrm{SL}_2(\mathbb{F}_q)$, Algebra and Applications 13,
DOI 10.1007/978-0-85729-157-8_12,

Lang's theorem. *If* $\mathbf{H}$ *is a* **connected** *algebraic group and if* $F\colon \mathbf{H} \to \mathbf{H}$ *is a Frobenius endomorphism of* $\mathbf{H}$, *then the morphism* $\mathbf{H} \to \mathbf{H}$, $h \mapsto h^{-1}F(h)$ *is an unramified Galois covering with group* $\mathbf{H}^F$; *in particular, it is surjective.*

12.1. Deligne-Lusztig Induction

If $w \in W$, we fix a representative $\dot{w}$ of w in $\mathrm{N}_{\mathbf{G}}(\mathbf{T})$ and denote by $wF\colon \mathbf{T} \to \mathbf{T}$, $t \mapsto \dot{w}F(t)\dot{w}^{-1}$. Then wF is a Frobenius endomorphism $\mathbf{T}$. The *Deligne-Lusztig variety* is the variety

$$\mathbf{Y}(\dot{w}) = \{g\mathbf{U} \in \mathbf{G}/\mathbf{U} \mid g^{-1}F(g) \in \mathbf{U}\dot{w}\mathbf{U}\}.$$

The action of $\mathbf{G}^F$ by left translations on $\mathbf{G}/\mathbf{U}$ stabilises $\mathbf{Y}(\dot{w})$ as does the action of $\mathbf{T}^{wF}$ by right translations on $\mathbf{G}/\mathbf{U}$ (recall that $\mathbf{T}$ normalises $\mathbf{U}$). Hence $\mathbf{Y}(\dot{w})$ comes equipped with the structure of a $(\mathbf{G}^F, \mathbf{T}^{wF})$-variety and its cohomology groups $H^i_c(\mathbf{Y}(\dot{w}))$ inherit the structure of a $(K\mathbf{G}^F, K\mathbf{T}^{wF})$-bimodule. We define

$$\begin{array}{rcl} \mathrm{R}_w\colon \mathscr{K}_0(K\mathbf{T}^{wF}) & \longrightarrow & \mathscr{K}_0(K\mathbf{G}^F) \\ {[M]}_{K\mathbf{T}^{wF}} & \longmapsto & \sum_{i \geqslant 0} (-1)^i \,[H^i_c(\mathbf{Y}(\dot{w})) \otimes_{K\mathbf{T}^{wF}} M]_{K\mathbf{G}^F} \end{array}$$

and

$$\begin{array}{rcl} {}^*\mathrm{R}_w\colon \mathscr{K}_0(K\mathbf{G}^F) & \longrightarrow & \mathscr{K}_0(K\mathbf{T}^{wF}) \\ {[M]}_{K\mathbf{G}^F} & \longmapsto & \sum_{i \geqslant 0} (-1)^i \,[H^i_c(\mathbf{Y}(\dot{w}))^* \otimes_{K\mathbf{G}^F} M]_{K\mathbf{G}^F} \end{array}$$

The morphisms R_w and ${}^*\mathrm{R}_w$ are called *Deligne-Lusztig induction* and *restriction* respectively. One may easily verify (thanks to Lang's Theorem, see Exercise 12.1) that these morphisms depend only on w and not on the choice of a representative $\dot{w}$. They are adjoint (for the standard scalar product on the Grothendieck groups).

If $w \in W$ and if θ is a linear character of $\mathbf{T}^{wF}$, the (virtual) character $\mathrm{R}_w(\theta)$ of $\mathbf{G}^F$ is called a *Deligne-Lusztig character*. One of the first fundamental results of this theory is the *Mackey formula* [DeLu, Theorem 6.8].

Mackey formula. *Let* w *and* w' *be two elements of* W *and let* θ *and* θ' *be two linear characters of* $\mathbf{T}^{wF}$ *and* $\mathbf{T}^{w'F}$ *respectively. Then*

(12.1.1) $$\langle \mathrm{R}_w(\theta), \mathrm{R}_{w'}(\theta')\rangle_{\mathbf{G}^F} = |\{x \in W \mid xwF(x)^{-1} = w' \text{ and } \theta' = \theta \circ x\}|.$$

REMARK – Note that two Deligne-Lusztig characters can be orthogonal and yet have common irreducible factors. Indeed, such characters are *virtual* characters. □

The next corollary follows from Mackey formula by computing the norm of the difference of the two virtual characters involved.

Corollary 12.1.2. *If x and w are two elements of W, then*

$$\mathrm{R}_{x^{-1}wF(x)} = \mathrm{R}_w \circ x^\wedge,$$

where $x^\wedge : (\mathbf{T}^{x^{-1}wF(x)})^\wedge \xrightarrow{\sim} (\mathbf{T}^{wF})^\wedge, \theta \mapsto \theta \circ x^{-1}$.

The second fundamental result is the following [DeLu, Corollary 7.7].

Theorem 12.1.3. *If γ is an irreducible character of $\mathbf{G}^F$, then there exists $w \in W$ and a linear character θ of $\mathbf{T}^{wF}$ such that $\langle \mathrm{R}_w(\theta), \gamma \rangle_{\mathbf{G}^F} \neq 0$.*

It follows that, in order to parametrise the irreducible characters of $\mathbf{G}^F$, it is "enough" to decompose the Deligne-Lusztig characters. This enormously difficult work was completed by Lusztig in 1984 (for groups with connected centre, which constitutes the largest part of the work). An important notion used to achieve this parametrisation is that of *geometric conjugacy*. In order to define this, let us introduce an integer $n_0 \geqslant 1$ such that the Frobenius endomorphisms $(\dot{w}F)^{n_0}$ and F^{n_0} of $\mathbf{G}$ agree for all $w \in W$ and induce the identify on W (such an n_0 always exists). Denote by

$$\begin{aligned} \mathrm{N}_w : \mathbf{T}^{F^{n_0}} &\longrightarrow \mathbf{T}^{wF} \\ t &\longmapsto t\, {}^{wF}t \cdots {}^{(wF)^{n_0-1}}t \end{aligned}$$

the norm map.

Let $\nabla(\mathbf{G}, F)$ denote the set of couples (w, θ), where $w \in W$, and θ is a linear character of $\mathbf{T}^{wF}$. If (w, θ) and (w', θ') are two elements of $\nabla(\mathbf{G}, F)$, we say that (w, θ) and (w', θ') are *geometrically conjugate* if the characters $\theta \circ \mathrm{N}_w$ and $\theta' \circ \mathrm{N}_{w'}$ of $\mathbf{T}^{F^{n_0}}$ are conjugate under W. If this is the case we write $(w, \theta) \approx (w', \theta')$. A *geometric series* is an equivalence class for the relation $\approx$. The proof of the following theorem may be found in [DeLu, Theorem 6.1].

Theorem 12.1.4. *Let (w, θ) and (w', θ') be two elements of $\nabla(\mathbf{G}, F)$ such that $\mathrm{R}_w(\theta)$ and $\mathrm{R}_{w'}(\theta')$ contain a common irreducible factor. Then $(w, \theta) \approx (w', \theta')$.*

If $\mathscr{S}$ is a geometric series, we set

$$\mathscr{E}(\mathbf{G}^F, \mathscr{S}) = \{\gamma \in \operatorname{Irr} \mathbf{G}^F \mid \exists\, (w, \theta) \in \mathscr{S},\ \langle \mathrm{R}_w(\theta), \gamma \rangle_{\mathbf{G}^F} \neq 0\}.$$

The set of characters $\mathscr{E}(\mathbf{G}^F, \mathscr{S})$ is called a *Lusztig series*. Combining Theorems 12.1.3 and 12.1.4, we obtain

$$\operatorname{Irr} \mathbf{G}^F = \dot{\bigcup_{\mathscr{S} \in \nabla(\mathbf{G}, F)/\approx}} \mathscr{E}(\mathbf{G}^F, \mathscr{S}). \tag{12.1.5}$$

Once one has obtained this initial description, "all that remains to do" is to parametrise the characters in a given Lusztig series. This was completed by Lusztig in a number of very long articles and a book [Lu2].

If $(w,\theta)\in\nabla(\mathbf{G},F)$, denote by $W(w,\theta)$ the stabiliser, in the group W, of the linear character $\theta\circ \mathrm{N}_w$ of $\mathbf{T}^{F^{n_0}}$. Then $W(w,\theta)$ is a wF-stable subgroup of W. If the centre of $\mathbf{G}$ is connected, then $W(w,\theta)$ is a subgroup of W generated by reflections.

Theorem 12.1.6 (Lusztig). *Let $\mathscr{S}$ be a geometric series and let $(w,\theta)\in\mathscr{S}$. If the centre of $\mathbf{G}$ is connected, then $\mathscr{E}(\mathbf{G}^F,\mathscr{S})$ is in bijection with a subset which only depends on the pair $(W(w,\theta),wF)$, and not on the group $\mathbf{G}$ or the cardinality q of the finite field.*

EXAMPLE 12.1.7 – If $W(w,\theta)=1$, then $|\mathscr{E}(\mathbf{G}^F,\mathscr{S})|=1$ and

$$\mathscr{E}(\mathbf{G}^F,\mathscr{S})=\{\varepsilon(w)\mathrm{R}_w(\theta)\},$$

where $\varepsilon(w)$ is the sign of w. □

Via the bijection of Theorem 12.1.6, Lusztig gives formulas for the degrees of the characters, for their multiplicities in all Deligne-Lusztig characters etc. He also deduces an analogous result for groups with non-connected centre, but for this one needs further techniques (see [Lu3]) which we will not recall here. In the group $G=\mathrm{SL}_2(\mathbb{F}_q)$ the non-connectedness of the centre is responsible, for example, both for the decomposition of the characters $\mathrm{R}(\alpha_0)$ and $\mathrm{R}'(\theta_0)$ and for the existence of quasi-isolated blocks.

We call a *unipotent character* of $\mathbf{G}^F$ any element of $\mathscr{E}(\mathbf{G}^F,\mathscr{S}_1)$, where $\mathscr{S}_1$ is the geometric series $\{(w,1)\mid w\in W\}$. In this case $W(w,1)=W$ and we conclude from Theorem 12.1.6 that

(12.1.8) $|\mathscr{E}(\mathbf{G}^F,\mathscr{S}_1)|$ *only depends on the pair (W,F), not on q.*

EXAMPLE 12.1.9 – Let us suppose that $\mathbf{G}=\mathrm{SL}_2(\mathbb{F})$ and that $F:\mathbf{G}\to\mathbf{G}$ is the natural split Frobenius endomorphism

$$F\begin{pmatrix} a & b\\ c & d\end{pmatrix}=\begin{pmatrix} a^q & b^q\\ c^q & d^q\end{pmatrix}.$$

Then $\mathbf{G}^F$ is our favourite finite group $G=\mathrm{SL}_2(\mathbb{F}_q)$. Moreover, $W=\{1,\bar{s}\}$, where $\bar{s}$ denotes the class of s in $N_{\mathbf{G}}(\mathbf{T})/\mathbf{T}$. We have

$$\mathbf{T}^F=T\quad\text{and}\quad \mathbf{Y}(1)=\mathbf{G}^F/\mathbf{U}^F=G/U$$

therefore $H_c^*(\mathbf{Y}(1))=K[\mathbf{G}^F/\mathbf{U}^F]$ by Theorem A.2.1(c), which shows that the Deligne-Lusztig induction R_1 is nothing but the Harish-Chandra induction R defined in Section 3.2.

Let us now turn to to the Deligne-Lusztig induction map associated to s. On the other hand, it is a well-known geometrical fact that the morphism of varieties

$$\begin{array}{ccc} \mathbf{G}/\mathbf{U} & \longrightarrow & \mathbf{A}^2(\mathbb{F}) \setminus \{(0,0)\} \\ \begin{pmatrix} a & b \\ c & d \end{pmatrix} & \longmapsto & (a,c) \end{array}$$

is an isomorphism. Via this isomorphism we have

$$\mathbf{T}^{sF} \simeq \mu_{q+1} \quad \text{and} \quad \mathbf{Y}(s) \simeq \mathbf{Y},$$

the isomorphism being compatible with the actions of G, μ_{q+1} and F. As a consequence,

$$\mathrm{R}_s = -\mathrm{R}'.$$

The map R' is therefore none other (up to a sign) than the Deligne-Lusztig induction associated to the non-trivial element of the Weyl group. The reader may verity that the above Mackey formula 12.1.1 is then a condensed version of the various Mackey formulas (3.2.4, 4.1.7 and 4.2.1) obtained for our group $\mathrm{SL}_2(\mathbb{F}_q)$. Moreover,

$$\mathscr{E}(G, \mathscr{S}_1) = \{1, \mathrm{St}_G\},$$

which allows one to verify 12.1.8.

Furthermore, $(1,1) \equiv (s,1)$, $(1,\alpha_0) \equiv (s,\theta_0)$ and

$$W(w,\theta) = \begin{cases} W & \text{if } \theta^2 = 1, \\ 1 & \text{if } \theta^2 \neq 1. \end{cases}$$

One may then verify the above claims concerning Lusztig series. □

We finish this section by recalling some geometric properties of the varieties $\mathbf{Y}(\dot{w})$.

Proposition 12.1.10. *Let $w \in W$. Then:*

(a) *The variety $\mathbf{Y}(\dot{w})$ is quasi-affine, smooth, and purely of dimension $l(w) = \dim \mathbf{B}w\mathbf{B}/\mathbf{B}$.*
(b) *The group $\mathbf{T}^{wF}$ acts freely on $\mathbf{Y}(\dot{w})$.*
(c) *The stabilisers, in $\mathbf{G}^F$, of elements of $\mathbf{Y}(\dot{w})$ are p-groups.*

REMARK – It is conjectured that Deligne-Lusztig varieties are always affine (and not only quasi-affine) but for the moment this is only known when q is large enough [DeLu, Theorem 9.7] and in certain particular cases. The smallest examples where it is not known if the varieties $\mathbf{Y}(\dot{w})$ are affine occur in the group of type G_2, when $q = 2$ and $l(w) = 3$. □

12.2. Modular Representations

Recall that

ℓ is a prime number different from p.

Because of its use of ℓ-adic cohomology and geometric methods, Deligne-Lusztig theory is particularly well adapted to the study of representations in unequal characteristic. Here we will recall some facts which illustrate this phenomenon.

12.2.1. Blocks

If $(w,\theta) \in \nabla(\mathbf{G},F)$, we denote by $\theta_{\ell'}$ the ℓ'-part of θ. A pair (w,θ) is called *ℓ-regular* if $\theta = \theta_{\ell'}$. We denote by $\nabla_{\ell'}(\mathbf{G},F)$ the set of ℓ-regular elements of $\nabla(\mathbf{G},F)$. If $\mathscr{S}$ is a geometric series, we denote by

$$\mathscr{S}_{\ell'} = \{(w,\theta_{\ell'}) \mid (w,\theta) \in \mathscr{S}\}.$$

Then $\mathscr{S}_{\ell'}$ is also a geometric series. A geometric series $\mathscr{S}$ is called *ℓ-regular* if $\mathscr{S} = \mathscr{S}_{\ell'}$. If $\mathscr{S}$ is an ℓ-regular geometric series, we denote by

$$\mathscr{E}_\ell(\mathbf{G}^F,\mathscr{S}) = \dot{\bigcup_{\substack{\mathscr{S}' \in \nabla(\mathbf{G},F)/\approx \\ \mathscr{S}'_{\ell'} = \mathscr{S}}}} \mathscr{E}(\mathbf{G}^F,\mathscr{S}').$$

The following result is due to Broué and Michel [BrMi, Theorem 2.2].

Theorem 12.2.1 (Broué-Michel). *If $\mathscr{S}$ is an ℓ-regular geometric series, then $\mathscr{E}_\ell(\mathbf{G}^F,\mathscr{S})$ is a union of ℓ-blocks.*

In other words, if we set

$$e_{\mathscr{S}}^{(\ell)} = \sum_{\chi \in \mathscr{E}_\ell(\mathbf{G}^F,\mathscr{S})} e_\chi,$$

then Theorem 12.2.1 of Broué and Michel may be translated as

(12.2.2) $$e_{\mathscr{S}}^{(\ell)} \in \mathscr{O}\mathbf{G}^F.$$

12.2.2. *Modular Deligne-Lusztig Induction*

If $w \in W$, we denote by $\mathscr{R}_w$ and ${}^*\mathscr{R}_w$ the functors of *Deligne-Lusztig induction* and *restriction:*

$$\begin{array}{rlcl} \mathscr{R}_w: & \mathsf{D}^b(\mathscr{O}\mathbf{T}^{wF}) & \longrightarrow & \mathsf{D}^b(\mathscr{O}\mathbf{G}^F) \\ & M & \longmapsto & \mathbf{R}\Gamma_c(\mathbf{Y}(\dot{w}), \mathscr{O}) \otimes_{\mathscr{O}\mathbf{T}^{wF}} M \end{array}$$

$$\begin{array}{rlcl} \mathscr{R}_w: & \mathsf{D}^b(\mathscr{O}\mathbf{G}^F) & \longrightarrow & \mathsf{D}^b(\mathscr{O}\mathbf{T}^{wF}) \\ & M & \longmapsto & \mathbf{R}\Gamma_c(\mathbf{Y}(\dot{w}), \mathscr{O})^* \otimes_{\mathscr{O}\mathbf{G}^F} M. \end{array}$$

These functors are well-defined and adjoint to one another. Indeed, the complex of bimodules $\mathbf{R}\Gamma_c(\mathbf{Y}(\dot{w}), \mathscr{O})$ is perfect as a left and right module by virtue of Proposition 12.1.10 and Theorem A.1.5.

If $\Lambda \in \{K, \mathscr{O}, k\}$, we denote by $\Lambda\mathscr{R}_w$ and $\Lambda^*\mathscr{R}_w$ the extension of scalars of the functors $\mathscr{R}_w$ and ${}^*\mathscr{R}_w$ respectively. For example, the functor $K\mathscr{R}_w$ (respectively $K^*\mathscr{R}_w$) induces the morphism R_w (respectively ${}^*\mathrm{R}_w$) between Grothendieck groups.

The Morita equivalence of Theorems 8.1.1 and 8.1.4 may be generalised to the case of arbitrary finite reductive groups in the following way [Bro, Theorem 3.3].

Theorem 12.2.3 (Broué). *Let* $(w, \theta) \in \nabla_{\ell'}(\mathbf{G}, F)$ *such that* $W(w, \theta) = 1$ *and let* $\mathscr{S}$ *be the geometric series which contains it. Then the cohomology group* $H_c^{l(w)}(\mathbf{Y}(\dot{w}), \mathscr{O})$ *induces a Morita equivalence between* $\mathscr{O}\mathbf{T}^{wF} e_\theta^{(\ell)}$ *and* $\mathscr{O}\mathbf{G}^F e_{\mathscr{S}}^{(\ell)}$.

In Theorem 12.2.3, $e_\theta^{(\ell)}$ denotes the primitive central idempotent of $\mathscr{O}\mathbf{T}^{wF}$ such that $\theta \in \operatorname{Irr} K\mathbf{T}^{wF} e_\theta^{(\ell)}$.

12.2.3. *The Geometric Version of Broué's Conjecture*

We will conclude this overview with a geometric version of Broué's conjecture (the general conjecture is given in Appendix B). For this, we make the following hypotheses, which allow us to considerably simplify the exposition.

Hypotheses and notation. *In this and only this subsection we suppose that:*

(1) *F acts trivially on W (if $\mathbf{G}$ is semi-simple, this is equivalent to saying that F is a split Frobenius endomorphism of $\mathbf{G}$).*
(2) *ℓ does not divide $|W|$.*
(3) *Denote by d the order of q modulo ℓ: we suppose that d is a **regular number** for W in the sense of Springer [Spr, §4].*

Broué's conjecture (geometric version). *There exists an element $w \in W$ of order d and regular in the sense of Springer (see [Spr, §4]), as well as a complex of $(\mathcal{O}\mathbf{G}^F, \mathcal{O}\mathrm{N}_{\mathbf{G}^{\dot{w}F}}(\mathbf{T}))$-bimodules $\mathscr{C}_w$ such that:*

(a) $\mathrm{Res}_{\mathbf{G}^F \times (\mathbf{T}^{wF})^{\mathrm{opp}}}^{\mathbf{G}^F \times (\mathrm{N}_{\mathbf{G}^{\dot{w}F}}(\mathbf{T}))^{\mathrm{opp}}} \mathscr{C}_w \simeq_{\mathrm{D}} \mathbf{R\Gamma}_c(\mathbf{Y}(\dot{w}), \mathcal{O})$.
(b) *$\mathscr{C}_w$ induces a Rickard equivalence between the principal blocks of $\mathbf{G}^F$ and $\mathrm{N}_{\mathbf{G}^{\dot{w}F}}(\mathbf{T})$.*

REMARK – Note that, under the previous assumptions, if S denotes a Sylow ℓ-subgroup of $\mathbf{G}^F$, then $\mathrm{N}_{\mathbf{G}^F}(S)$ is isomorphic to $\mathrm{N}_{\mathbf{G}^{\dot{w}F}}(\mathbf{T})$. □

This version has been shown in very few cases. The case where ℓ divides $q-1$ was shown by Puig (see, for example, [CaEn, Theorem 23.12]):

Theorem 12.2.4 (Puig). *If $d = 1$ (that is to say if ℓ divides $q-1$), then the geometric version of Broué's conjecture is true. In this case the induced equivalence is even a Morita equivalence, because the Deligne-Lusztig variety in question is of dimension zero.*

The next result was proved by Rouquier and the author [BoRo, Theorem 4.6].

Theorem 12.2.5. *If $\mathbf{G}^F = \mathrm{GL}_n(\mathbb{F}_q)$, $\mathrm{SL}_n(\mathbb{F}_q)$ or $\mathrm{PGL}_n(\mathbb{F}_q)$ and if $d = n$, then the geometric version of Broué's conjecture is true.*

EXAMPLE – In the case where $\mathbf{G}^F = \mathrm{SL}_2(\mathbb{F}_q)$, we have already seen Puig's theorem (see Corollary 8.3.3) and Theorem 12.2.5 (see Corollary 8.3.7). □

Very recently, Olivier Dudas has shown in his thesis [Du] the final result of this chapter.

Theorem 12.2.6 (Dudas). *If $\mathbf{G}$ is of type B_n, C_n and $d = 2n$, then the geometric version of Broué's conjecture is true.*

Exercises

12.1. Let $\dot{w}$ and $\ddot{w}$ be two representatives of the same element $w \in W$ in $N_{\mathbf{G}}(\mathbf{T})$. Show that the $(\mathbf{G}^F, \mathbf{T}^{wF})$-varieties $\mathbf{Y}(\dot{w})$ and $\mathbf{Y}(\ddot{w})$ are isomorphic. Deduce that the morphisms R_w and $^*\mathrm{R}_w$ (as well as the functors $\mathscr{R}_w$ and $^*\mathscr{R}_w$) do not depend on a choice of representative $\dot{w}$ of w. (***Hint:*** Use Lang's theorem.)

12.2. Verify the assertions of Example 12.1.9. Use the Lang map $\mathbf{G} \to \mathbf{G}$, $g \mapsto g^{-1}F(g)$ and Lang's theorem to show that $\mathbf{Y}/G \simeq \mathbf{A}^1(\mathbb{F})$.

Appendix A
ℓ-Adic Cohomology

The étale topology, the fundamentals of which were developed in the *Séminaire de Géométrie Algébrique du Bois-Marie* in the 1960's (see, for example, [SGA1], [SGA4], [SGA4$\frac{1}{2}$]) allows one to associate functorially to any algebraic variety a complex and corresponding cohomology groups (with coefficients in Λ, which can be K, $\mathcal{O}$ or $\mathcal{O}/\mathfrak{l}^n$, for $n \geqslant 1$) enjoying numerous remarkable properties. In this appendix we recall those properties which are essential to our goal of studying the representations of $\mathrm{SL}_2(\mathbb{F}_q)$ (functoriality, perfectness, exact sequences, bounds, Künneth formula, Poincaré duality, traces, the Lefschetz fixed-point theorem, ...). We refer the reader to [SGA4$\frac{1}{2}$] for all results stated without reference.

A.1. Properties of the Complex*

The complex of ℓ-adic cohomology inherits an action of the monoid of endomorphisms of the variety. Even better, it can be realised as a complex of modules for this monoid (see the works of Rickard [Ric3] and Rouquier [Rou3]).

Therefore, in this appendix we work under the following hypotheses. Let $\mathbf{V}$ be a quasi-projective algebraic variety defined over $\mathbb{F}$ and let Γ be a *monoid* acting on $\mathbf{V}$ via endomorphisms. We also set Λ to be one of K, $\mathcal{O}$ or $\mathcal{O}/\mathfrak{l}^n$ (for $n \geqslant 1$).

EXAMPLE – We can take $\mathbf{V} = \mathbf{Y}$ (the Drinfeld curve) and $\Gamma = G \times (\mu_{q+1} \rtimes \langle F \rangle_{\mathrm{mon}})$. □

By [Rou3], there exists a bounded complex $\mathbf{R\Gamma}_c(\mathbf{V}, \Lambda)$ of $\Lambda\Gamma$-modules whose cohomology groups, which we denote $H^i_c(\mathbf{V}, \Lambda)$, are, as $\Lambda\Gamma$-modules, the cohomology groups with compact support of the variety $\mathbf{V}$ (with coefficients in the constant sheaf Λ). To simplify notation we denote by $H^i_c(\mathbf{V})$ the $K\Gamma$-module $H^i_c(\mathbf{V}, K)$.

C. Bonnafé, *Representations of* $\mathrm{SL}_2(\mathbb{F}_q)$, Algebra and Applications 13,
DOI 10.1007/978-0-85729-157-8,

Recall that

(A.1.1) $$\mathbf{R\Gamma}_c(\mathbf{V},\Lambda) \simeq \Lambda \otimes^{\mathsf{L}}_{\mathcal{O}} \mathbf{R\Gamma}_c(\mathbf{V},\mathcal{O}).$$

Here, $\otimes^{\mathsf{L}}_{\mathcal{O}}$ denotes the derived functor of the tensor product. In particular, as K is $\mathcal{O}$-flat,

(A.1.2) $$\mathbf{R\Gamma}_c(\mathbf{V}) \simeq K \otimes_{\mathcal{O}} \mathbf{R\Gamma}_c(\mathbf{V},\mathcal{O})$$

and

(A.1.3) $$H^i_c(\mathbf{V},K) \simeq K \otimes_{\mathcal{O}} H^i_c(\mathbf{V},\mathcal{O}).$$

Moreover, by [SGA4$\frac{1}{2}$, Arcata, III, §3],

(A.1.4) *if $\mathbf{V}$ is a smooth **curve**, then $H^i_c(\mathbf{V},\mathcal{O})$ is torsion-free.*

Another essential property is given in the following proposition.

Theorem A.1.5 (Rickard, Rouquier). *Suppose that Γ is a finite group. Then:*

(a) $\mathbf{R\Gamma}_c(\mathbf{V},\Lambda)$ *is homotopic to a bounded complex of $\Lambda\Gamma$-modules of finite type.*
(b) *If the stabilisers of points of $\mathbf{V}$ are of order invertible in Λ, then $\mathbf{R\Gamma}_c(\mathbf{V},\Lambda)$ is homotopic to a bounded complex of projective $\Lambda\Gamma$-modules.*

A.2. Properties of the Cohomology Groups

A.2.1. General Properties

We begin by recalling certain properties of cohomology groups with coefficients in the general ring Λ.

Theorem A.2.1. *Let $d = \dim \mathbf{V}$ and let $\mathscr{I}(\mathbf{V})$ denote the set of irreducible components of $\mathbf{V}$ of dimension d. Then:*

(a) $H^i_c(\mathbf{V},\Lambda)$ *is a Λ-module of finite type.*
(b) $H^i_c(\mathbf{V},\Lambda) = 0$ *if $i < 0$ or if $i > 2d$. If $\mathbf{V}$ is affine and purely of dimension d, then, moreover, $H^i_c(\mathbf{V}) = 0$ if $i < d$.*
(c) *If Γ is a group we have an isomorphism of $\Lambda\Gamma$-modules $H^{2d}_c(\mathbf{V}) \simeq \Lambda[\mathscr{I}(\mathbf{V})]$.*
(d) *If $\mathbf{U}$ is a Γ-stable open subvariety of $\mathbf{V}$ with closed complement $\mathbf{Z}$, then we have a long exact sequence of $\Lambda\Gamma$-modules*

$$\cdots \longrightarrow H^i_c(\mathbf{U},\Lambda) \longrightarrow H^i_c(\mathbf{V},\Lambda) \longrightarrow H^i_c(\mathbf{Z},\Lambda) \longrightarrow H^{i+1}_c(\mathbf{U},\Lambda) \longrightarrow \cdots$$

(e) *If Γ is contained in a **connected** algebraic group acting **regularly** on $\mathbf{V}$, then Γ acts trivially on $H^i_c(\mathbf{V},\Lambda)$.*

(f) $H^i_c(\mathbf{A}^d(\mathbb{F}),\Lambda) = \begin{cases} \Lambda & \text{if } i = 2d, \\ 0 & \text{otherwise.} \end{cases}$

Proof. (a), (b), (c), (d) and (f) are shown in [SGA4$\frac{1}{2}$]. (e) is shown in [DeLu, Proposition 6.4]. □

A.2.2. Cohomology with Coefficients in K

We will now state some properties which hold when $\Lambda = K$.

Künneth formula. *If $\mathbf{V}'$ is another variety on which Γ acts, then we have an isomorphism of $K\Gamma$-modules*

(A.2.2)
$$H^r_c(\mathbf{V}\times\mathbf{V}') \simeq \bigoplus_{i=0}^{r} H^i_c(\mathbf{V}) \otimes_K H^{r-i}_c(\mathbf{V}')$$

for all $r \geqslant 0$.

If Δ is a finite subgroup of Γ which is normalised by Γ (i.e. $\gamma\Delta = \Delta\gamma$ for all $\gamma \in \Gamma$), then the quotient monoid Γ/Δ acts on the quotient variety $\mathbf{V}/\Delta$ and we have an isomorphism of $\Lambda(\Gamma/\Delta)$-modules

(A.2.3)
$$H^i_c(\mathbf{V}/\Delta) \simeq H^i_c(\mathbf{V})^{\Delta}.$$

Poincaré duality. *If $\mathbf{V}$ is **irreducible, projective** and **smooth** of dimension d, then we have, for all $i \in \{0,1,2,\dots,2d\}$, a perfect Γ-equivariant duality*

(A.2.4)
$$H^i_c(\mathbf{V}) \times H^{2d-i}_c(\mathbf{V}) \longrightarrow H^{2d}_c(\mathbf{V}).$$

REMARK – Note that, under the above hypotheses, $\dim_K H^{2d}_c(\mathbf{V}) = 1$ (see Theorem A.2.1 (c)). However, it is possible for Γ not to act trivially on $H^{2d}_c(\mathbf{V})$. For example, a Frobenius endomorphism over $\mathbb{F}_q$ acts as multiplication by q^d (see Theorem A.2.7(b) below). □

A.2.3. The Euler Characteristic

It is often useful to consider the *Euler characteristic* of $\mathbf{V}$. If one takes into account the action of the monoid Γ, the Euler characteristic becomes not just a number, but an element in the Grothendieck group $\mathcal{K}_0(K\Gamma)$ of the category of $K\Gamma$-modules *of finite dimension over* K defined by

$$H^*_c(\mathbf{V}) = \sum_{i\geqslant 0} (-1)^i\, [H^i_c(\mathbf{V})]_{\Gamma}.$$

We use the notation $H_c^*(\mathbf{V})_\Gamma$ if we wish to emphasise the monoid Γ.

If $\gamma \in \mathrm{End}(\mathbf{V})$ or Γ, we denote by $\mathrm{Tr}^*_{\mathbf{V}}(\gamma)$ the alternating sum

$$\mathrm{Tr}^*_{\mathbf{V}}(\gamma) = \sum_{i \geqslant 0} (-1)^i \,\mathrm{Tr}(\gamma, H_c^i(\mathbf{V})).$$

The function $\mathrm{Tr}^*_{\mathbf{V}}\colon \mathrm{End}(\mathbf{V}) \longrightarrow K$ is called the *Lefschetz character* of $\mathbf{V}$. Recall that, if γ is an automorphism of $\mathbf{V}$ of finite order, then

(A.2.5) $$\mathrm{Tr}^*_{\mathbf{V}}(\gamma) \in \mathbb{Z}.$$

Theorem A.2.6. *Let $d = \dim \mathbf{V}$. Then:*

(a) *If $\mathbf{U}$ is an open Γ-stable subvariety of $\mathbf{V}$ and if $\mathbf{Z}$ denotes its closed complement, then $H_c^*(\mathbf{V}) = H_c^*(\mathbf{U}) + H_c^*(\mathbf{Z})$. In particular, $\mathrm{Tr}^*_{\mathbf{V}} = \mathrm{Tr}^*_{\mathbf{U}} + \mathrm{Tr}^*_{\mathbf{Z}}$.*

(b) *If Δ is a finite subgroup of Γ normalised by Γ, then $H_c^*(\mathbf{V}/\Delta)_{\Gamma/\Delta} = H_c^*(\mathbf{V})^\Delta$.*

(c) *If $\mathbf{V}'$ is an algebraic variety upon which Γ also acts, then $H_c^*(\mathbf{V} \times \mathbf{V}') = H_c^*(\mathbf{V}) \otimes H_c^*(\mathbf{V}')$.*

(d) *If s and u are two invertible elements of Γ such that $su = us$, s is of order prime to p and u is of order a power of p, then*

$$\mathrm{Tr}^*_{\mathbf{V}}(su) = \mathrm{Tr}^*_{\mathbf{V}^s}(u).$$

(e) *If Γ is a finite group and $\mathbf{S}$ is a torus acting on $\mathbf{V}$ and commuting with the action of Γ, then $H_c^*(\mathbf{V}) = H_c^*(\mathbf{V}^{\mathbf{S}})$.*

Proof. (a), (b) and (c) may be found in [SGA4$\frac{1}{2}$], while (d) is shown in [DeLu, Theorem 3.2] and (e) is shown, for example, in [DiMi, Proposition 10.15]. □

A.2.4. Action of a Frobenius Endomorphism

Suppose that $\mathbf{V}$ is defined over the finite field $\mathbb{F}_q$, with associated Frobenius endomorphism F. Then F acts on the cohomology groups $H_c^i(\mathbf{V})$. We assemble here some classical results concerning this action.

Theorem A.2.7. *We have:*

(a) $\mathrm{Tr}^*_{\mathbf{V}}(F) = |\mathbf{V}^F|$ *(Lefschetz fixed-point theorem).*

(b) *If $\mathbf{V}$ is irreducible of dimension d, then F acts on $H_c^{2d}(\mathbf{V})$ as multiplication by q^d.*

(c) *The eigenvalues of F on $H_c^i(\mathbf{V})$ are algebraic integers of the form $\omega q^{j/2}$, where j is a natural number such that $0 \leqslant j \leqslant i$, and ω is an algebraic number, all of whose complex conjugates are of norm 1. In particular, F is an automorphism of the K-vector space $H_c^i(\mathbf{V})$.*

(d) *If $\mathbf{V}$ is **projective** and **smooth**, then the eigenvalues of F on $H_c^i(\mathbf{V})$ are algebraic integers, all of whose complex conjugates have norm $q^{i/2}$.*

Proof. (a) and (b) are shown in [SGA4$\frac{1}{2}$]. (c) and (d) have been shown twice by Deligne [De1], [De2]: these results constitute the last difficulty in resolving the celebrated *Weil conjectures*, which are analogues for algebraic varieties of the Riemann hypothesis for number fields. □

A.3. Examples

A.3.1. The Projective Line

We identify $\mathbf{A}^1(\mathbb{F})$ with the open subvariety of $\mathbf{P}^1(\mathbb{F})$ equal to $\{[x;y] \in \mathbf{P}^1(\mathbb{F}) \mid y \neq 0\}$ and set $\infty = [1;0]$. By Theorem A.2.1(d), we have an exact sequence

$$\begin{aligned} 0 \longrightarrow H^0_c(\mathbf{A}^1(\mathbb{F}),\Lambda) \longrightarrow H^0_c(\mathbf{P}^1(\mathbb{F}),\Lambda) \longrightarrow H^0_c(\infty,\Lambda) \longrightarrow \\ H^1_c(\mathbf{A}^1(\mathbb{F}),\Lambda) \longrightarrow H^1_c(\mathbf{P}^1(\mathbb{F}),\Lambda) \longrightarrow H^1_c(\infty,\Lambda) \longrightarrow \\ H^2_c(\mathbf{A}^1(\mathbb{F}),\Lambda) \longrightarrow H^2_c(\mathbf{P}^1(\mathbb{F}),\Lambda) \longrightarrow H^2_c(\infty,\Lambda) \longrightarrow 0. \end{aligned}$$

Using Theorem A.2.1(f) for $d = 0$ and $d = 1$, we obtain an exact sequence

$$\begin{aligned} 0 \longrightarrow 0 \longrightarrow H^0_c(\mathbf{P}^1(\mathbb{F}),\Lambda) \longrightarrow \Lambda \longrightarrow \\ 0 \longrightarrow H^1_c(\mathbf{P}^1(\mathbb{F}),\Lambda) \longrightarrow 0 \longrightarrow \\ \Lambda \longrightarrow H^2_c(\mathbf{P}^1(\mathbb{F}),\Lambda) \longrightarrow 0 \longrightarrow 0. \end{aligned}$$

We conclude:

$$H^i_c(\mathbf{P}^1(\mathbb{F}),\Lambda) = \begin{cases} \Lambda & \text{if } i = 0 \text{ or } 2, \\ 0 & \text{otherwise.} \end{cases} \tag{A.3.1}$$

On the other hand, the connected group $\mathrm{GL}_2(\mathbb{F})$ acts regularly on $\mathbf{P}^1(\mathbb{F})$, therefore, by Theorem A.2.1(e),

(A.3.2) $\mathrm{GL}_2(\mathbb{F})$ *acts trivially on* $H^i_c(\mathbf{P}^1(\mathbb{F}))$.

To conclude, we can keep track of the action of the Frobenius endomorphism $F\colon \mathbf{P}^1(\mathbb{F}) \longrightarrow \mathbf{P}^1(\mathbb{F}), [x;y] \mapsto [x^q;y^q]$ in the above exact sequences. Using Theorem A.2.7(b), we obtain

(A.3.3) $F = 1$ *on* $H^0_c(\mathbf{P}^1(\mathbb{F}))$ *and* $F = q$ *on* $H^2_c(\mathbf{P}^1(\mathbb{F}))$.

In particular, we have verified the Weil conjectures (Theorem A.2.7(c)) and the Lefschetz fixed-point theorem (Theorem A.2.7(a)) which says that

$$|\mathbf{P}^1(\mathbb{F}_q)| = q+1,$$

which is no surprise.

A.3.2. The One-Dimensional Torus

Denote by $\mathbf{U} = \mathbf{A}^1(\mathbb{F}) \setminus \{0\}$. By Theorem A.2.1(d), we have an exact sequence

$$\begin{aligned} 0 \longrightarrow H_c^0(\mathbf{U},\Lambda) &\longrightarrow H_c^0(\mathbf{A}^1(\mathbb{F}),\Lambda) \longrightarrow H_c^0(0,\Lambda) \longrightarrow \\ H_c^1(\mathbf{U},\Lambda) &\longrightarrow H_c^1(\mathbf{A}^1(\mathbb{F}),\Lambda) \longrightarrow H_c^1(0,\Lambda) \longrightarrow \\ H_c^2(\mathbf{U},\Lambda) &\longrightarrow H_c^2(\mathbf{A}^1(\mathbb{F}),\Lambda) \longrightarrow H_c^2(0,\Lambda) \longrightarrow 0. \end{aligned}$$

Using Theorem A.2.1(f) for $d = 0$ and $d = 1$, we therefore obtain an exact sequence

$$\begin{aligned} 0 \longrightarrow H_c^0(\mathbf{U},\Lambda) &\longrightarrow 0 \longrightarrow \Lambda \longrightarrow \\ H_c^1(\mathbf{U},\Lambda) &\longrightarrow 0 \longrightarrow 0 \longrightarrow \\ H_c^2(\mathbf{U},\Lambda) &\longrightarrow \Lambda \longrightarrow 0 \longrightarrow 0. \end{aligned}$$

We conclude:

$$H_c^i(\mathbf{A}^1(\mathbb{F}) \setminus \{0\},\Lambda) = \begin{cases} \Lambda & \text{if } i = 1 \text{ or } 2, \\ 0 & \text{otherwise.} \end{cases} \tag{A.3.4}$$

On the other hand, the connected group $\mathbb{F}^\times$ acts regularly (by multiplication) on $\mathbf{A}^1(\mathbb{F}) \setminus \{0\}$, therefore, by Theorem A.2.1(e),

$$\mathbb{F}^\times \textit{ acts trivially on } H_c^i(\mathbf{A}^1(\mathbb{F}) \setminus \{0\}). \tag{A.3.5}$$

To conclude, we can keep track of the action of the Frobenius endomorphism $F\colon \mathbf{A}^1(\mathbb{F}) \longrightarrow \mathbf{A}^1(\mathbb{F})$, $x \mapsto x^q$ in the above exact sequence. Using Theorem A.2.7(b), we obtain

$$F = 1 \textit{ on } H_c^1(\mathbf{A}^1(\mathbb{F}) \setminus \{0\}) \quad \textit{and} \quad F = q \textit{ on } H_c^2(\mathbf{A}^1(\mathbb{F}) \setminus \{0\}). \tag{A.3.6}$$

The Lefschetz fixed-point theorem (Theorem A.2.7(a)) tells us that

$$|\mathbf{A}^1(\mathbb{F}_q) \setminus \{0\}| = q-1,$$

which is again no surprise.

Exercises

A.1. Determine the cohomology of $\mathbf{P}^n(\mathbb{F})$.

A.2. Let $a_1, \ldots, a_n$ be pairwise distinct points of $\mathbf{A}^1(\mathbb{F})$. Calculate the cohomology of $\mathbf{A}^1(\mathbb{F}) \setminus \{a_1, \ldots, a_n\}$.

A.3. In statement (c) of Theorem A.2.7, the algebraic number ω is not necessarily an algebraic integer. This may occur and when it does it considerably restricts the values of ω. Indeed, show that an *algebraic integer* (over $\mathbb{Z}$) all of whose complex conjugates are of norm 1 is a root of unity.

Appendix B
Block Theory

Fix a finite group Γ. In this appendix we recall the "essential" results (that is, essential for our purposes) of block theory, where the object of study is the representations of the algebras $\mathcal{O}\Gamma$ and $k\Gamma$. As in the rest of this book we assume that the algebras $K\Gamma'$ and $k\Gamma'$ are split for all groups Γ' met in this appendix. We denote by $\mathcal{O}\Gamma \to k\Gamma$, $a \mapsto \bar{a}$ the reduction modulo $\mathfrak{l}$, and extend this notation to $\mathcal{O}$-modules.

For further details and developments the reader may consult one of the many references on this subject: [Alp], [CuRe, Chapters 2 and 7], [Isa, Chapter 15], [NaTs], [Ser, Part 3], [The].

B.1. Definition

Set $\Lambda = \mathcal{O}$ or $\Lambda = k$. A Λ-*bloc* of Γ is an indecomposable direct summand of the $(\Lambda\Gamma, \Lambda\Gamma)$-bimodule $\Lambda\Gamma$. If we decompose the $(\Lambda\Gamma, \Lambda\Gamma)$-bimodule $\Lambda\Gamma$ as a direct sum of indecomposable factors in two different ways

$$\Lambda\Gamma = A_1 \oplus \cdots \oplus A_r = A'_1 \oplus \cdots \oplus A'_s,$$

and write $1 = e_1 + \cdots + e_r = e'_1 + \cdots + e'_s$, where $e_i \in A_i$ and $e'_j \in A'_j$. Then, the next proposition follows from [The, Corollary 4.2].

Proposition B.1.1. *With the above notation we have:*

(a) $r = s$ *and there exists a permutation* σ *of* $\{1, 2, \ldots, r\}$ *such that* $A'_i = A_{\sigma(i)}$ *for all* i *(and therefore* $e'_i = e_{\sigma(i)}$*).*
(b) e_i *is a primitive central idempotent of* $\Lambda\Gamma$ *and* $A_i = \Lambda\Gamma e_i$*; furthermore,* A_i *is a* Λ*-algebra with identity* e_i*.*
(c) *If* $i \neq j$*, then* $e_i e_j = e_j e_i = 0$.
(d) *The* Λ*-algebras* $\Lambda\Gamma$ *and* $A_1 \times \cdots \times A_r$ *are isomorphic.*

C. Bonnafé, *Representations of* $SL_2(\mathbb{F}_q)$, Algebra and Applications 13,
DOI 10.1007/978-0-85729-157-8, © Springer-Verlag London Limited 2011

The previous proposition shows that the set of Λ-blocks of Γ is well-defined. The following proposition (which follows from results about lifting idempotents [The, Theorems 3.1 and 3.2]) shows that it does not depend too much on Λ.

Proposition B.1.2. *If A is an $\mathcal{O}$-block of Γ, then $\overline{A}$ is a k-block of Γ. If e is a primitive central idempotent of $\mathcal{O}\Gamma$, then $\overline{e}$ is a primitive central idempotent of $k\Gamma$.*

The reduction modulo $\mathfrak{l}$ induces a bijection between the $\mathcal{O}$-blocks and k-blocks of Γ. It also induces a bijection between the primitive central idempotents of $\mathcal{O}\Gamma$ and of $k\Gamma$.

The isomorphism of $\mathcal{O}$-algebras $\mathcal{O}\Gamma \simeq A_1 \times \cdots \times A_r$ of Proposition B.1.1 induces an isomorphism $K\Gamma \simeq KA_1 \times \cdots \times KA_r$ and hence induces a partition

$$\operatorname{Irr} K\Gamma = \operatorname{Irr}(KA_1) \,\dot{\cup}\, \cdots \,\dot{\cup}\, \operatorname{Irr}(KA_r). \tag{B.1.3}$$

We call an *ℓ-block* of Γ a subset of $\operatorname{Irr} K\Gamma$ of the form $\operatorname{Irr}(KA)$, where A is an $\mathcal{O}$-block of Γ. If χ is an irreducible character of Γ and if $a \in \mathrm{Z}(\mathcal{O}\Gamma)$, we denote by $\omega_\chi(a)$ the scalar by which a acts on an irreducible representation affording χ as character (by virtue of Schur's lemma). Then $\omega_\chi : \mathrm{Z}(\mathcal{O}\Gamma) \to \mathcal{O}$ is a morphism of $\mathcal{O}$-algebras which will be called the *central character* associated to χ. If X is a subset of Γ, we set $\hat{X} = \sum_{\gamma \in X} \gamma \in \mathcal{O}\Gamma$. Recall that $(\widehat{\mathrm{Cl}_\Gamma(\gamma)})_{\gamma \in [\Gamma/\sim]}$ is a $\mathcal{O}$-basis of $\mathrm{Z}(\mathcal{O}\Gamma)$. Moreover, recall [Isa, Theorem 3.7] that

$$\omega_\chi(\widehat{\mathrm{Cl}_\Gamma(\gamma)}) = \frac{\chi(\gamma)\,|\mathrm{Cl}_\Gamma(\gamma)|}{\chi(1)} \in \mathcal{O}. \tag{B.1.4}$$

The following proposition gives a characterisation of the partition into ℓ-blocks of $\operatorname{Irr} K\Gamma$ using central characters (see, for example, [Isa, Theorem 15.18]).

Proposition B.1.5. *Let χ and χ' be two irreducible characters of Γ. Then the following are equivalent:*

(1) *χ and χ' are in the same ℓ-block.*
(2) $\omega_\chi(a) \equiv \omega_{\chi'}(a) \mod \mathfrak{l}$ *for all $a \in \mathrm{Z}(\mathcal{O}\Gamma)$.*
(3) $\dfrac{\chi(\gamma)\,|\mathrm{Cl}_\Gamma(\gamma)|}{\chi(1)} \equiv \dfrac{\chi'(\gamma)\,|\mathrm{Cl}_\Gamma(\gamma)|}{\chi'(1)} \mod \mathfrak{l}$ *for all $\gamma \in \Gamma$.*

B.2. Brauer Correspondents

B.2.1. Brauer's Theorems

If D is a subgroup of Γ, we define

$$\begin{aligned}\mathrm{Br}_D: \quad Z(k\Gamma) &\longrightarrow Z(kN_\Gamma(D))\\ \sum_{\gamma\in\Gamma} a_\gamma\gamma &\longmapsto \sum_{\gamma\in C_\Gamma(D)} a_\gamma\gamma.\end{aligned}$$

Then [Isa, Lemma 15.32] we have the following.

Proposition B.2.1. *If D is an **ℓ-subgroup** of Γ, then Br_D is a morphism of k-algebras.*

When D is an ℓ-subgroup of Γ, Br_D is called the *Brauer morphism.* If A is an $\mathcal{O}$-block, we call the *defect group* of A any ℓ-subgroup D of Γ such that $\mathrm{Br}_D(\overline{A}) \neq 0$ and which is maximal with this property. Then [Isa, Lemma 15.33].

Proposition B.2.2. *If D and D' are two defect groups of an $\mathcal{O}$-block of Γ, then D and D' are conjugate in Γ.*

Proposition B.2.2 justifies the abuse of language committed by referring to *the* defect group of A. The following proposition relates the ℓ-valuation of the degree of a character to that of the order of the defect group of the corresponding ℓ-block [Isa, Theorem 15.41].

Proposition B.2.3. *If D is the defect group of A and if $\chi \in \mathrm{Irr}(KA)$, then*

$$\frac{|D|\cdot\chi(1)}{|\Gamma|} \in \mathcal{O}.$$

Moreover, there exists at least one character $\chi \in \mathrm{Irr}(KA)$ such that

$$\frac{|D|\cdot\chi(1)}{|\Gamma|} \in \mathcal{O}^\times.$$

EXAMPLE B.2.4 – Let χ an irreducible character of Γ such that $|\Gamma|/\chi(1) \in \mathcal{O}^\times$. Then $e_\chi \in Z(\mathcal{O}\Gamma)$ and therefore e_χ is a primitive central idempotent of $\mathcal{O}\Gamma$ (because it is already primitive in $K\Gamma$). As a consequence, χ is alone in its ℓ-block. Moreover, Proposition B.2.3 shows that the defect group of $\mathcal{O}\Gamma e_\chi$ is trivial. □

For a proof of the following theorem the reader is referred to [The, Corollary 37.13] or [Isa, Theorem 15.45].

Brauer's first main theorem. *Let D be an ℓ-subgroup of Γ. The map Br_D induces a bijection between the set of $\mathcal{O}$-blocks of Γ with defect group D and the set of $\mathcal{O}$-blocks of $N_\Gamma(D)$ with defect group D.*

If A is an $\mathcal{O}$-block of Γ with defect group D and if $A^\flat$ is the $\mathcal{O}$-block of $N_\Gamma(D)$ associated to A by the bijection of Brauer's first main theorem, we say that $A^\flat$ is the *Brauer correspondent* of A. It is characterised as follows: $A^\flat$ is the unique $\mathcal{O}$-block of $N_\Gamma(D)$ such that $\mathrm{Br}_D(\overline{A})\overline{A^\flat} \neq 0$. If we denote by e (respectively $e^\flat$) the primitive central idempotent of $\mathcal{O}\Gamma$ (respectively $\mathcal{O}N_\Gamma(D)$)

such that $A = \mathcal{O}\Gamma e$ (respectively $A^\flat = \mathcal{O}\mathrm{N}_\Gamma(D)e^\flat$), then $e^\flat$ is characterised by the property that $\mathrm{Br}_D(\overline{e}) = \overline{e}^\flat$.

We call the *principal block* of Γ the unique $\mathcal{O}$-block A of Γ such that $1_\Gamma \in \mathrm{Irr}(KA)$. We denote the principal block of Γ by $B_0(\Gamma)$.

Brauer's third main theorem. *Let S be a Sylow ℓ-subgroup of Γ. Then S is the defect group of $B_0(\Gamma)$ and $B_0(\mathrm{N}_\Gamma(S))$ is the Brauer correspondent of $B_0(\Gamma)$.*

B.2.2. Conjectures

One of the central problems in block theory is to relate the representations in an $\mathcal{O}$-block to those of its Brauer correspondent. There exist diverse conjectures in this direction. We will be content state the following.

McKay conjecture (global). *Denote by $\mathrm{Irr}_{\ell'}(\Gamma)$ the set of characters of Γ of degree prime to ℓ and let S be a Sylow ℓ-subgroup of Γ. Then*

$$|\mathrm{Irr}_{\ell'}(K\Gamma)| = |\mathrm{Irr}_{\ell'}(K\mathrm{N}_\Gamma(S))|.$$

In case the defect group is abelian, Broué proposed the following conjecture. It is of a much more structural nature, but there does not exist a version if the defect group is not abelian.

Broué's conjecture. *Let A be a block of Γ with defect group D and let $A^\flat$ be its Brauer correspondent. If D is abelian, then the derived categories $\mathrm{D}^b(A)$ and $\mathrm{D}^b(A^\flat)$ are equivalent as triangulated categories.*

This is not the place to enter into an exhaustive description of the numerous variants and refinements of these conjectures. We remark however that if the Sylow ℓ-subgroup of Γ is abelian, then Broué's conjecture implies McKay's conjecture. Note also that these conjectures (and their variants) have been verified in a very large number of cases.

B.2.3. Equivalences of Categories: Methods

In order to obtain Morita equivalences or derived equivalences for our group $\mathrm{SL}_2(\mathbb{F}_q)$ we will make use of some general results. Fix A and A' as two sums of $\mathcal{O}$-blocks of finite groups Γ and Γ'. A fundamental property of blocks of finite groups is that these algebras are *symmetric* [The, §6]. This fact will considerably simplify our work in what follows.

Morita equivalences. Fix an (A, A')-bimodule M. We will recall some criteria used to verify that the functor $M \otimes_{A'} - : A'\text{-mod} \longrightarrow A\text{-mod}$ is an equivalence of categories (then called a *Morita equivalence* between A and A'). The following theorem is shown in [Bro, Theorem 0.2].

Theorem B.2.5 (Broué). *Suppose that the bimodule M is projective both as a left A-module and as a right A'-module. Then the following properties are equivalent:*

(1) *The functor $M \otimes_{A'} - : A'\text{-mod} \longrightarrow A\text{-mod}$ is a Morita equivalence.*
(2) *The functor $KM \otimes_{KA'} - : KA'\text{-mod} \longrightarrow KA\text{-mod}$ is a Morita equivalence.*
(3) *Every irreducible character of KA is an irreducible component of the KG-module KM, and the map $\operatorname{Irr} KA' \to \operatorname{Irr} KA$, $[V]_{KA'} \mapsto [KM \otimes_{KA'} V]_{KA}$ is well-defined and bijective.*
(4) *Every irreducible character of KA is a factor of KM and the natural map $KA' \to \operatorname{End}_{KA}(KM)^{\mathrm{opp}}$ is an isomorphism.*
(5) *Every irreducible character of KA is a factor of KM and the natural map $A' \to \operatorname{End}_A(M)^{\mathrm{opp}}$ is an isomorphism.*

Rickard equivalences. Let $\mathscr{C}$ be a complex of (A, A')-bimodules. We recall some criteria which may be used to verify that the functor

$$\mathscr{C} \otimes^{\mathsf{L}}_{A'} - : \mathsf{D}^b(A') \longrightarrow \mathsf{D}^b(A)$$

is an equivalence of categories (then called a *Rickard equivalence* between A and A'). Recall first that, if $\mathscr{C} = M[i]$ (that is to say, the complex with one non-zero term M in degree $-i$) and if M induces a Morita equivalence between A and A', then $\mathscr{C}$ induces a Rickard equivalence between A and A'.

We denote by $H^\bullet(\mathscr{C})$ the (A, A')-bimodule $\oplus_{i\in\mathbb{Z}} H^i(\mathscr{C})$. Note that, as K is $\mathscr{O}$-flat, we have

$$H^\bullet(K\mathscr{C}) = KH^\bullet(\mathscr{C}).$$

The following theorem is part of the folklore of the subject, but we have not been able to find a satisfactory reference (note however, that the proof below is very strongly influenced by [Ric4, Theorem 2.1]).

Theorem B.2.6. *Suppose that $\mathscr{C}$ is perfect, both as a complex of left A-modules and of right A'-modules, and that*

$$\operatorname{Hom}_{\mathsf{D}^b(A)}(\mathscr{C}, \mathscr{C}[i]) = 0$$

for all $i \neq 0$. Then the following properties are equivalent:

(1) *The functor $\mathscr{C} \otimes_{A'} -$ is a Rickard equivalence between A and A'.*
(2) *All irreducible characters of KA are factors of $H^\bullet(K\mathscr{C})$ and the natural map $A' \to \operatorname{End}_{\mathsf{D}^b(A)}(\mathscr{C})^{\mathrm{opp}}$ is an isomorphism.*

Proof. Firstly, it is clear that (1) implies (2). It remains to show that (2) implies (1). Therefore suppose that all irreducible characters of KA occur as factors of $H^\bullet(K\mathscr{C})$ and that the natural map $A' \to \operatorname{End}_{\mathsf{D}^b(A)}(\mathscr{C})^{\mathrm{opp}}$ is an isomorphism.

First step: reduction to the case where A is an $\mathscr{O}$-block. Let e be the central idempotent of $\mathscr{O}\Gamma$ such that $A = \mathscr{O}\Gamma e$ and let $e = e_1 + \cdots + e_n$ be a decomposition of

e into a sum of primitive central idempotents. Set $A_i = \mathcal{O}\Gamma e_i = Ae_i$. Then the action of e_i on $\mathscr{C}$ is an element of $\mathrm{End}_{\mathrm{D}^b(A)}(\mathscr{C})$, therefore there exists an element e'_i of A' such that the endomorphism induced by right multiplication by e'_i is equal (in the derived category) to that induced by left multiplication by e_i. Moreover, e'_i is idempotent and central (because this endomorphism commutes with the action of A'). Let $A'_i = A'e'_i$. Then

$$A = \prod_{i=1}^{n} A_i, \quad A' = \prod_{i=1}^{n} A'_i \quad \text{and} \quad \mathscr{C} = \bigoplus_{i=1}^{n} e_i\mathscr{C} = \bigoplus_{i=1}^{n} \mathscr{C}e'_i.$$

It is now sufficient to show that $e_i\mathscr{C} = \mathscr{C}e'_i$ induces a Rickard equivalence between A_i and A'_i. It is easy to verify that $e_i\mathscr{C}$ satisfies the same hypotheses as $\mathscr{C}$ and so we can (and will) suppose that A is an $\mathcal{O}$-block of $\mathcal{O}\Gamma$.

Second step: adjunction. Firstly, after replacing $\mathscr{C}$ by a homotopic complex, we may suppose that $\mathscr{C}$ is a complex of (A, A')-bimodules which are projective as left and right modules. As the algebras A and A' are symmetric, we have isomorphisms

$$\mathscr{C}^* \simeq_{\mathrm{C}} \mathrm{Hom}_{\mathrm{C}^b(A)}(\mathscr{C}, A) \simeq_{\mathrm{C}} \mathrm{Hom}_{\mathrm{C}^b(A')}(\mathscr{C}, A').$$

As a consequence, the functors $\mathscr{C} \otimes_{A'} -$ and $\mathscr{C}^* \otimes_A -$ *between the categories of complexes* $\mathrm{C}^b(A)$ *and* $\mathrm{C}^b(A')$ are left and right adjoint to each other. The unit and counit of these adjunctions induce natural morphisms

$$A \xrightarrow{\alpha} \mathscr{C} \otimes_A \mathscr{C}^* \xrightarrow{\beta} A$$

of complexes of (A, A)-bimodules. Tensoring with $\mathscr{C}$ we obtain natural morphisms of complexes of (A, A')-bimodules

$$\mathscr{C} \xrightarrow{\alpha'} \mathscr{C} \otimes_A \mathscr{C}^* \otimes_A \mathscr{C} \xrightarrow{\beta'} \mathscr{C}.$$

It follows from general properties of adjoint functors that α' (respectively β') is a split injection (respectively split surjection) [McL, Theorem IV.1.1]. We may therefore write $\mathscr{C} \otimes_A \mathscr{C}^* \otimes_A \mathscr{C} \simeq \mathscr{C} \oplus \mathscr{C}_0$, where $\mathscr{C}_0$ is a complex of (A, A')-bimodules projective as left and right modules. By assumption, the natural morphism $A' \to \mathscr{C}^* \otimes_A \mathscr{C}$ is an isomorphism in the derived category (that is a quasi-isomorphism) and therefore $\mathscr{C} \otimes_A \mathscr{C}^* \otimes_A \mathscr{C} \simeq_{\mathrm{D}} \mathscr{C}$. As a consequence, $\mathscr{C}_0$ is acyclic and α' and β' are quasi-isomorphisms. In particular, $\beta' \circ \alpha'$ is a quasi-isomorphism.

On the other hand, $\beta \circ \alpha$ is an endomorphism of the (A, A)-bimodule A and therefore there exists an element z in the centre $\mathrm{Z}(A)$ of A such that $\beta \circ \alpha(a) = za$ for all $a \in A$. Suppose that $\beta \circ \alpha$ is not an isomorphism. Then z is an element in the radical of $\mathrm{Z}(A)$ (as A is a block, $\mathrm{Z}(A)$ is a local ring). In this case $\beta' \circ \alpha'$ cannot be a quasi-isomorphism (as $\mathscr{C}$ is non-zero by hypothesis). We have therefore shown that $\beta \circ \alpha$ is an isomorphism.

Third step: conclusion. We may therefore write $\mathscr{C} \otimes_A \mathscr{C}^* = A \oplus \mathscr{C}_1$, where $\mathscr{C}_1$ is a complex of (A, A)-bimodules, projective as left and right modules. Now, we have an isomorphism (in the derived category)

$$\mathscr{C} \otimes_A \mathscr{C}^* \otimes_{A'} \mathscr{C} \otimes_A \mathscr{C}^* \simeq_{\mathrm{D}} \mathscr{C} \otimes_A A \otimes_A \mathscr{C}^* \simeq_{\mathrm{D}} \mathscr{C} \otimes_A \mathscr{C}^*.$$

Therefore

$$A \simeq_{\mathrm{D}} A \oplus \mathscr{C}_1 \oplus \mathscr{C}_1 \oplus \mathscr{C}_1 \otimes_A \mathscr{C}_1.$$

Therefore $\mathscr{C}_1$ is acyclic, which shows that

$$\mathscr{C} \otimes_A \mathscr{C}^* \simeq_{\mathrm{D}} A$$

and completes the proof of the theorem. □

REMARK B.2.7 – If $\mathscr{C}$ induces a Rickard equivalence between A and A', then:

(a) The complex $k\mathscr{C}$ induces a Rickard equivalence between kA and kA'.
(b) The complex $K\mathscr{C}$ induces a Rickard equivalence between KA and KA'.
(c) The centres of A and A' are isomorphic.
(d) $|\operatorname{Irr} KA| = |\operatorname{Irr} KA'|$ and $|\operatorname{Irr} kA| = |\operatorname{Irr} kA'|$. □

B.3. Decomposition Matrices

We begin with a classical result in representation theory [CuRe, Propositions 23.16 and 16.16].

Proposition B.3.1. *Let R be a principal ideal domain and A an R-algebra which is of finite type and free as an R-module. Denote by $\mathbb{K}$ the field of fractions of R. Let V be a $\mathbb{K}A$-module of finite type. Then there exists an A-stable R-lattice M of V (so that $V = \mathbb{K} \otimes_R V$). Moreover, for all maximal ideals $\mathfrak{m}$ of R, the class $[M/\mathfrak{m}M]_{A/\mathfrak{m}A}$ in the Grothendieck group $\mathscr{K}_0(A/\mathfrak{m}A)$ does not depend on the choice of the A-submodule M of V such that $V = \mathbb{K} \otimes_R M$.*

If A is an $\mathscr{O}$-algebra which is free and of finite type as an $\mathscr{O}$-module, Proposition B.3.1 shows that we can define a *decomposition map*

$$\mathrm{dec}_A \colon \mathscr{K}_0(KA) \longrightarrow \mathscr{K}_0(kA)$$

as follows: if V is a KA-module, we may find in V an A-stable $\mathscr{O}$-lattice M (so that $V = KM$) and we set

$$\mathrm{dec}_A([V]_{KA}) = [kM]_{kA}.$$

We denote by $\mathrm{Dec}(A)$ the *decomposition matrix* of A, that is to say, the transpose of the matrix of the map dec_A in the canonical bases $([V]_{KA})_{V \in \operatorname{Irr} KA}$ and

$([S]_{kA})_{S\in \mathrm{Irr}\, kA}$. Its rows are indexed by $\mathrm{Irr}\, KA$, while its columns are indexed by $\mathrm{Irr}\, kA$.

In the case of group algebras and their blocks, we recall the following results (see [The, Theorems 42.3 and 42.8] or [Isa, Corollary 2.7 and 15.11]).

Proposition B.3.2. *We have:*

(a) $|\mathrm{Irr}\, K\Gamma|$ *is equal to the number of conjugacy classes of elements of* Γ.
(b) $|\mathrm{Irr}\, k\Gamma|$ *is equal to the number of conjugacy classes of* ℓ*-regular elements of* Γ.
(c) $\mathrm{dec}_{\mathcal{O}\Gamma} = \bigoplus_{A\in\{\mathcal{O}\text{-blocks of }\Gamma\}} \mathrm{dec}_A$ *and* $\mathrm{Dec}(\mathcal{O}\Gamma) = \bigoplus_{A\in\{\mathcal{O}\text{-blocks of }\Gamma\}} \mathrm{Dec}(A)$.

EXAMPLE B.3.3 – If A is an $\mathcal{O}$-block of Γ with trivial defect group and if V denotes the unique (up to isomorphism) simple KA-module (see Example B.2.4), then, for all A-stable $\mathcal{O}$-lattices M of V, kM is the unique simple kA-module. As a consequence, $\mathrm{Dec}(A) = (1)$.

In particular, if Γ is an ℓ'-group, then all $\mathcal{O}$-blocks of Γ have trivial defect group and hence the decomposition map $\mathrm{dec}_{\mathcal{O}\Gamma}$ induces a bijection between $\mathrm{Irr}\, K\Gamma$ and $\mathrm{Irr}\, k\Gamma$ and the decomposition matrix $\mathrm{Dec}(\mathcal{O}\Gamma)$ is the identity. □

B.4. Brauer Trees*

Hypothesis. *We fix an* $\mathcal{O}$*-block* A *of* Γ *with defect group* D*. In this section only we suppose that the group* D *is* ***cyclic***.

For the definitions and results referred to without reference in this section, we refer the reader to [HiLu]. To each block of a group with cyclic defect group is associated a graph, called the *Brauer tree* (the graph is in fact a tree). We recall its construction.

Let $A^\flat$ denote the Brauer correspondent of A. The primitive central idempotent $e^\flat$ of $k\mathrm{N}_\Gamma(D)$ such that $\overline{A}^\flat = k\mathrm{N}_\Gamma(D)e^\flat$ is in fact an element of $k\mathrm{C}_\Gamma(D)$. We denote by d the number of primitive central idempotents of $k\mathrm{C}_\Gamma(D)$ occurring in the decomposition of $e^\flat$. Recall that d divides $p-1$. The set $\mathrm{Irr}\, KA$ decomposes as follows:

$$\mathrm{Irr}\, KA = \{\chi_1, \ldots, \chi_d\} \,\dot\cup\, \{\chi_\lambda \mid \lambda \in \Lambda\},$$

where the χ_λ ($\lambda \in \Lambda$) are the *exceptional characters* of KA. We then set $\chi_{\mathrm{exc}} = \sum_{\lambda\in\Lambda} \chi_\lambda$ and

$$\mathcal{V}_A = \{\chi_1, \ldots, \chi_d, \chi_{\mathrm{exc}}\}.$$

The Brauer tree $\mathcal{T}_A$ of A is then the *pointed* graph (that is a graph with a distinguished vertex, called the *exceptional* vertex) defined as follows:

- The set of vertices of $\mathscr{T}_A$ is $\mathscr{V}_A$. The exceptional vertex is χ_{exc}.
- We join two distinct elements χ and χ' in $\mathscr{V}_A$ if $\chi + \chi'$ is the character of a projective indecomposable A-module.

The following facts are classical:

(a) $\mathscr{T}_A$ is a connected tree.
(b) If ψ is the character of an indecomposable projective A-module, then there exists two distinct elements χ and χ' of $\mathscr{V}_A$ such that $\psi = \chi + \chi'$. In particular, the isomorphism classes of projective indecomposable A-modules are in bijection with the edges of the Brauer tree $\mathscr{T}_A$.

In depictions of the Brauer trees given in this book the exceptional vertex will be represented by a round black circle ● and the non-exceptional vertices will be represented by a round white circle ○.

REMARK B.4.1 – The Brauer tree of a general finite group is equipped with an extra piece of data, its *planar embedding*, that is to say its representation in the plane. This data is fundamental, but requires much more information to obtain. As almost all the Brauer trees which occur in this book have only one planar embedding we will not preoccupy ourselves with this issue. □

A fundamental theorem in the theory of blocks of groups with cyclic defect group is the following.

Theorem B.4.2 (Brauer). *Let Γ and Γ' be two finite groups and let A and A' be two $\mathcal{O}$-blocks with cyclic defect groups of the same order. Then A and A' are Morita equivalent if and only if the Brauer trees $\mathscr{T}_A$ and $\mathscr{T}_{A'}$ are isomorphic (as pointed graphs equipped with a planar embedding).*

EXAMPLE B.4.3 – Suppose that $\Gamma = \mathfrak{A}_5$, the alternating group of degree 5, and that $\ell = 5$. Recall that $\operatorname{Irr} \mathfrak{A}_5 = \{\chi_1, \chi_3, \chi_3', \chi_4, \chi_5\}$, where the index denotes the degree of the character. If A the principal $\mathcal{O}$-block of Γ, then $\operatorname{Irr} KA = \{\chi_1, \chi_3, \chi_3', \chi_4\}$, with defect group $D =< (1,2,3,4,5) >$ a Sylow 5-subgroup of Γ and $\mathscr{V}_A = \{\chi_1, \chi_4, \chi_3 + \chi_3'\}$. The Brauer tree $\mathscr{T}_A$ is therefore

$\mathscr{T}_A$ 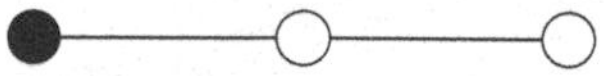

On the other hand, the reader may verify that the Brauer tree of the principal block $A^\flat$ of the normaliser of D in Γ (that is to say of the Brauer correspondent of A) is

$\mathscr{T}_{A^\flat}$ 

where $\operatorname{Irr} KA^\flat = \{1, \varepsilon, \psi_2, \psi_2'\}$, and 1 and ε (respectively ψ_2 and ψ_2') are the two characters of $KA^\flat$ of degree 1 (respectively 2). The characters of degree 2 are the exceptional characters.

By virtue of Brauer's Theorem B.4.2, the blocks A and $A^\flat$ are not Morita equivalent: in fact, $\mathscr{T}_A$ and $\mathscr{T}_{A^\flat}$ are not isomorphic as **pointed** graphs, even though they are as graphs.

However Broué's conjecture, as verified by Rickard [Ric2], asserts that the derived categories $\mathsf{D}^b(A)$ and $\mathsf{D}^b(A^\flat)$ are equivalent as triangulated categories. □

Appendix C
Review of Reflection Groups

Fix a field $\mathbb{K}$ of characteristic *zero* as well as a $\mathbb{K}$-vector space V of finite dimension n. If $g \in \mathrm{GL}_{\mathbb{K}}(V)$, we say that g is a *reflection* if $\mathrm{Ker}(g - \mathrm{Id}_V)$ is of codimension 1. If Γ is a finite subgroup of $\mathrm{GL}_{\mathbb{K}}(V)$, we denote by $\mathrm{Ref}(\Gamma)$ the set of its reflections. We say that Γ is a *reflection group* on V if it is generated by the reflections that it contains.

The $\mathbb{K}$-algebra of polynomial functions on V will be denoted $\mathbb{K}[V]$. This algebra inherits an action of $\mathrm{GL}_{\mathbb{K}}(V)$. A $\mathbb{K}$-algebra of finite type is called *polynomial* if it is isomorphic to an algebra $\mathbb{K}[V']$, where V' is a finite dimensional $\mathbb{K}$-vector space. If Γ is a finite subgroup of $\mathrm{GL}_n(\mathbb{K})$, the algebra of invariants $\mathbb{K}[V]^\Gamma$ is a $\mathbb{K}$-algebra of finite type. The Shephard-Todd-Chevalley theorem characterises the reflection groups in terms of their algebras of invariants [Bou, chapitre V, §5].

Shephard-Todd-Chevalley theorem. *Let Γ be a finite subgroup of $\mathrm{GL}_{\mathbb{K}}(V)$. Then Γ is a reflection group if and only if the invariant algebra $\mathbb{K}[V]^\Gamma$ is polynomial. If this is the case, then there exists n homogeneous and algebraically independent polynomials $p_1, \dots, p_n \in \mathbb{K}[V]$ such that $\mathbb{K}[V]^\Gamma = \mathbb{K}[p_1, \dots, p_n]$. If $d_i = \deg(p_i)$ and if the p_i are chosen so that $d_1 \leqslant d_2 \leqslant \cdots \leqslant d_n$, then:*

(a) $\displaystyle \frac{1}{|\Gamma|} \sum_{\gamma \in \Gamma} \frac{\mathrm{Tr}(\gamma)}{\det(1 - \gamma X)} = \prod_{i=1}^{n} \frac{1}{1 - X^{d_i}}$ *(here, X is an indeterminate).*

(b) *The sequence $(d_1, \dots, d_n)$ does not depend on the choice of the p_i. We call it the* ***sequence of degrees*** *of Γ.*

(c) $|\Gamma| = d_1 d_2 \cdots d_n$ *and* $|\mathrm{Ref}(\Gamma)| = \sum_{i=1}^{n} (d_i - 1)$.

(d) *If moreover Γ is an irreducible subgroup of $\mathrm{GL}_{\mathbb{K}}(V)$, then it is absolutely irreducible, its centre is cyclic and $|\mathrm{Z}(\Gamma)| = \mathrm{pgcd}(d_1, d_2, \dots, d_n)$.*

C. Bonnafé, *Representations of* $\mathrm{SL}_2(\mathbb{F}_q)$, Algebra and Applications 13,
DOI 10.1007/978-0-85729-157-8, © Springer-Verlag London Limited 2011

References

[Abh] S. ABHYANKAR, Coverings of algebraic curves, *Amer. J. Math.* **79** (1957), 825–856.

[Alp] J.L. ALPERIN, *Local representation theory*, Cambridge Studies in Advanced Mathematics **11**, Cambridge University Press, Cambridge, 1986, x + 178pp.

[Bor] A. BOREL, *Linear algebraic groups*, deuxième édition, Graduate Texts in Mathematics **126**, Springer-Verlag, New York, 1991, xii + 288pp.

[Bon] C. BONNAFÉ, Quasi-isolated elements in reductive groups, *Comm. Algebra* **33** (2005), 2315–2337.

[BoRo] C. BONNAFÉ & R. ROUQUIER, Coxeter orbits and modular representations, *Nagoya J. Math.* **183** (2006), 1–34.

[Bou] N. BOURBAKI, *Groupes et algèbres de Lie, chapitres IV, V et VI*, Hermann, Paris, 1968.

[Bro] M. BROUÉ, Isométries de caractères et équivalences de Morita ou dérivées, *Publ. Math. I.H.É.S.* **71** (1990), 45–63.

[BrMaMi] M. BROUÉ, G. MALLE & J. MICHEL, *Représentations unipotentes génériques des groupes réductifs finis*, Astérisque **212** (1993).

[BrMi] M. BROUÉ & J. MICHEL, Blocs et séries de Lusztig dans un groupe réductif fini, *J. Reine Angew. Math.* **395** (1989), 56–67.

[BrPu] M. BROUÉ & L. PUIG, A Frobenius theorem for blocks, *Invent. Math.* **56** (1980), 117–128.

[Ca] M. CABANES, On Okuyama's theorems about Alvis-Curtis duality, preprint (2008), arXiv:0810.0952.

[CaEn] M. CABANES & M. ENGUEHARD, *Representation theory of finite reductive groups*, New Mathematical Monographs **1**, Cambridge University Press, Cambridge, 2004, xviii + 436pp.

[CaRi] M. CABANES & J. RICKARD, Alvis-Curtis duality as an equivalence of derived categories, *Modular representation theory of finite groups (Charlottesville, VA, 1998)*, 157–174, de Gruyter, Berlin, 2001.

[Carter] R.W. CARTER, *Finite groups of Lie type. Conjugacy classes and complex characters*, Pure and Applied Mathematics, John Wiley & Sons (New York), 1985, xii + 544pp.

[Cartier] P. CARTIER, Représentations linéaires des groupes algébriques semi-simples en caractéristique non nulle, *Séminaire Bourbaki*, Vol. **8**, Exp. No. 255, 179–188, Soc. Math. France, Paris, 1995.

[CuRe] C.W. CURTIS & I. REINER, *Methods of representation theory, with applications to finite groups and orders*, vol. I, réédition de l'original de 1981, Wiley Classics Library, New York, 1990. xxiv + 819pp. *Methods of representation theory, with*

DOI 10.1007/978-0-85729-157-8, © Springer-Verlag London Limited 2011

applications to finite groups and orders, vol. II, réédition de l'original de 1987, Wiley Classics Library, New York, 1994. xviii + 951pp.

[De1] P. DELIGNE, La conjecture de Weil I, *Publ. Math. I.H.É.S.* **43** (1974), 273–307.

[De2] P. DELIGNE, La conjecture de Weil II, *Publ. Math. I.H.É.S.* **52** (1980), 137–252.

[DeLu] P. DELIGNE & G. LUSZTIG, Representations of reductive groups over finite fields, *Ann. of Math.* **103** (1976), 103–161.

[DiMi] F. DIGNE & J. MICHEL, *Representations of finite groups of Lie type*, London Mathematical Society Student Texts **21**, Cambridge University Press, 1991, iv + 159pp.

[Dri] V.G. DRINFELD, Elliptic modules (Russian), *Mat. Sb. (N.S.)* **94** (1974), 594–627.

[Du] O. DUDAS, *Géométrie des variétés de Deligne-Lusztig, décompositions, cohomologie modulo ℓ et représentations modulaires*, Ph. D. thesis, Université de Franche-Comté, Besançon, 2010.

[Go] B. GONARD, Catégories dérivées de blocs à défaut non abélien de $\mathsf{GL}_2(\mathbb{F}_q)$, Ph.D. Thesis, Université Paris VII (2002).

[HiLu] G. HISS & K. LUX, *Brauer trees of sporadic groups*, Oxford Science Publications, The Clarendon Press, Oxford University Press, New York, 1989, x + 526pp.

[Jan] J.C. JANTZEN, *Representations of algebraic groups, Second edition*, Mathematical Surveys and Monographs, **107**, American Mathematical Society, Providence, RI, 2003, xiv + 576pp.

[Jor] H.E. JORDAN, Group-characters of various types of linear groups, *Amer. J. Math.* **29** (1907), 387–405.

[Har] R. HARTSHORNE, *Algebraic geometry*, Graduate Texts in Mathematics **52**, Springer-Verlag, New York-Heidelberg, 1977, xvi + 496pp.

[Isa] M. I. ISAACS, *Character theory of finite groups*, AMS Chelsea Publishing, Providence, RI, 2006, xii + 310pp.

[Leh] G.I. LEHRER & D.E. TAYLOR, *Unitary reflection groups*, Cambridge University Press, Australian Mathematical Society Lecture Note Series **20**.

[Lin] M. LINCKELMANN, Derived equivalence for cyclic blocks over a p-adic ring, *Math. Z.* **207** (1991), 293–304.

[Lu1] G. LUSZTIG, *Representations of finite Chevalley groups*, CBMS Regional Conference (Madison 1977), CBMS Regional Conference Series in Mathematics **39**, American Mathematical Society, Providence, 1978, v + 48pp.

[Lu2] G. LUSZTIG, *Characters of reductive groups over finite fields*, Annals of Math. Studies **107** (1984), Princeton University Press, xxi + 384pp.

[Lu3] G. LUSZTIG, On the representations of reductive groups with non-connected center, *Astérisque* **168** (1988), 157–166.

[McL] S. MACLANE, *Categories for the working mathematician*, Graduate Texts in Mathematics **5**, Springer-Verlag, New York-Berlin, 1971, ix + 262pp.

[NaTs] H. NAGAO & Y. TSUSHIMA, *Representations of finite groups*, Academic Press, Inc., Boston, MA, 1989, xviii + 424pp.

[Oku1] T. OKUYAMA, Derived equivalences in $\mathsf{SL}(2, q)$, preprint (2000).

[Oku2] T. OKUYAMA, Remarks on splendid tilting complexes, dans *Representation theory of finite groups and related topics* (Kyoto, 1998), Surikaisekikenkyusho Kokyuroku **1149** (2000), 53–59.

[Oku3] T. OKUYAMA, On conjectures on complexes of some module categories related to Coxeter complexes, preprint (2006), 25pp.

[Ray] M. RAYNAUD, Revêtements de la droite affine en caractéristique $p > 0$ et conjecture d'Abhyankar, *Invent. Math.* **116** (1994), 425–462.

[Ric1] J. RICKARD, Derived categories and stable equivalence, *J. Pure and App. Alg.* **61** (1989), 303–317.

[Ric2] J. RICKARD, Derived equivalences for the principal blocks of $\mathfrak{A}_4$ and $\mathfrak{A}_5$, preprint (1990).

[Ric3] J. RICKARD, Finite group actions and étale cohomology, *Publ. Math. I.H.E.S* **80** (1994), 81–94.

[Ric4] J. RICKARD, Splendid equivalences: derived categories and permutation modules, *Proc. L.M.S.* **72** (1996), 331-358.

[Rou1] R. ROUQUIER, Some examples of Rickard complexes, Proceedings of the Conference on representation theory of groups, algebras, orders, *Ann. St. Univ. Ovidius Constantza* **4** (1996), 169–173.

[Rou2] R. ROUQUIER, The derived category of blocks with cyclic defect groups, dans *Derived equivalences for group rings*, 199–220, Lecture Notes in Math. **1685**, Springer, Berlin, 1998.

[Rou3] R. ROUQUIER, Complexes de chaînes étales et courbes de Deligne-Lusztig, *J. Algebra* **257** (2002), 482–508.

[Sch] I. SCHUR, Untersuchungen über die Darstellung des endlichen Gruppen durch geborchene lineare Substitutionen, *J. für Math.* **132** (1907).

[Ser] J.-P. SERRE, *Représentations linéaires des groupes finis*, Troisième édition, Hermann, Paris, 1978, 182pp. **English translation:** *Linear representations of finite groups*, Graduate Texts in Mathematics **42**, Springer-Verlag, New York-Heidelberg, 1977, x + 170pp.

[SGA1] A. GROTHENDIECK, *Revêtements étales et groupe fondamental (SGA1)*, Lecture Notes in Mathematics **224**, Springer, 1971.

[SGA4] M. ARTIN, A. GROTHENDIECK ET J.-L. VERDIER, *Théorie des topos et cohomologie étale des schémas (SGA4)*, Lecture Notes in Mathematics **269, 270, 305**, Springer, 1972–1973.

[SGA4$\frac{1}{2}$] P. DELIGNE, *Cohomologie étale (SGA4$\frac{1}{2}$)*, Lecture Notes in Mathematics **569**, Springer, 1977.

[ShTo] G.C. SHEPHARD & J.A. TODD, Finite unitary reflection groups, *Canadian J. Math.* **6** (1954), 274–304.

[Spr] T.A. SPRINGER, Regular elements of finite reflection groups, *Invent. Math.* **25** (1974), 159–198.

[The] J. THÉVENAZ, *G-algebras and modular representation theory*, Oxford Mathematical Monographs, Oxford Science Publications, The Clarendon Press, Oxford University Press, New York, 1995, xxviii + 470pp.

[We] A. WEIL, Numbers of solutions of equations in finite fields, *Bull. Amer. Math. Soc.* **55** (1949), 497–508.

[Yo] Y. YOSHII, Broué's conjecture for the nonprincipal block of $\mathsf{SL}(2,q)$ with full defect, *J. Algebra* **321** (2009), 2486–2499.

Index

Symbols

C. Bonnafé, *Representations of* $\mathrm{SL}_2(\mathbb{F}_q)$, Algebra and Applications 13,
DOI 10.1007/978-0-85729-157-8, © Springer-Verlag London Limited 2011